Voltage-Based Alternative Repair Criteria

A Report to the
Advisory Committee on Reactor Safeguards
by the
Ad Hoc Subcommittee on a
Differing Professional Opinion

Advisory Committee on Reactor Safeguards
U.S. Nuclear Regulatory Commission
Washington, DC 20555-0001

AVAILABILITY OF REFERENCE MATERIALS
IN NRC PUBLICATIONS

NRC Reference Material

As of November 1999, you may electronically access NUREG-series publications and other NRC records at NRC's Public Electronic Reading Room at www.nrc.gov/NRC/ADAMS/index.html.
Publicly released records include, to name a few, NUREG-series publications; *Federal Register* notices; applicant, licensee, and vendor documents and correspondence; NRC correspondence and internal memoranda; bulletins and information notices; inspection and investigative reports; licensee event reports; and Commission papers and their attachments.

NRC publications in the NUREG series, NRC regulations, and *Title 10, Energy*, in the Code of *Federal Regulations* may also be purchased from one of these two sources.
1. The Superintendent of Documents
   U.S. Government Printing Office
   Mail Stop SSOP
   Washington, DC 20402–0001
   Internet: bookstore.gpo.gov
   Telephone: 202-512-1800
   Fax: 202-512-2250
2. The National Technical Information Service
   Springfield, VA 22161–0002
   www.ntis.gov
   1–800–553–6847 or, locally, 703–605–6000

A single copy of each NRC draft report for comment is available free, to the extent of supply, upon written request as follows:
Address: Office of the Chief Information Officer,
              Reproduction and Distribution
                Services Section
              U.S. Nuclear Regulatory Commission
              Washington, DC 20555-0001
E-mail: DISTRIBUTION@nrc.gov
Facsimile: 301–415–2289

Some publications in the NUREG series that are posted at NRC's Web site address www.nrc.gov/NRC/NUREGS/indexnum.html are updated periodically and may differ from the last printed version. Although references to material found on a Web site bear the date the material was accessed, the material available on the date cited may subsequently be removed from the site.

Non-NRC Reference Material

Documents available from public and special technical libraries include all open literature items, such as books, journal articles, and transactions, *Federal Register* notices, Federal and State legislation, and congressional reports. Such documents as theses, dissertations, foreign reports and translations, and non-NRC conference proceedings may be purchased from their sponsoring organization.

Copies of industry codes and standards used in a substantive manner in the NRC regulatory process are maintained at—
         The NRC Technical Library
         Two White Flint North
         11545 Rockville Pike
         Rockville, MD 20852–2738

These standards are available in the library for reference use by the public. Codes and standards are usually copyrighted and may be purchased from the originating organization or, if they are American National Standards, from—
         American National Standards Institute
         11 West 42nd Street
         New York, NY 10036–8002
         www.ansi.org
         212–642–4900

Legally binding regulatory requirements are stated only in laws; NRC regulations; licenses, including technical specifications; or orders, not in NUREG-series publications. The views expressed in contractor-prepared publications in this series are not necessarily those of the NRC.

The NUREG series comprises (1) technical and administrative reports and books prepared by the staff (NUREG–XXXX) or agency contractors (NUREG/CR–XXXX), (2) proceedings of conferences (NUREG/CP–XXXX), (3) reports resulting from international agreements (NUREG/IA–XXXX), (4) brochures (NUREG/BR–XXXX), and (5) compilations of legal decisions and orders of the Commission and Atomic and Safety Licensing Boards and of Directors' decisions under Section 2.206 of NRC's regulations (NUREG–0750).

NUREG-1740

Voltage-Based Alternative Repair Criteria

A Report to the Advisory Committee on Reactor Safeguards
by the
Ad Hoc Subcommittee on a Differing Professional Opinion

Date Completed: February 2001
Date Published: March 2001

Advisory Committee on Reactor Safeguards
U.S. Nuclear Regulatory Commission
Washington, DC 20555-0001

ABSTRACT

This report was prepared for the Advisory Committee on Reactor Safeguards (ACRS) as part of the Committee's effort to provide comments and recommendations to the Executive Director for Operations of the U. S. Nuclear Regulatory Commission for use in resolving a differing professional opinion (DPO) concerning voltage-based alternative criteria for the repair of flaws in steam generator tubes in pressurized water reactors. The report was prepared by an Ad Hoc Subcommittee of ACRS and its consultants. The report discusses the contentions that have been raised in the DPO and the staff responses to these contentions. Analyses and experimental results that support the various positions are described in summary fashion. Based on this information, the Subcommittee reaches a variety of conclusions and recommendations. The Subcommittee finds that alternative repair criteria are needed and that the general features of the criteria and the condition monitoring program that the staff has endorsed provide such alternative repair criteria that can adequately protect public health and safety. Analyses of the risk associated with adoption of the repair criteria need to better consider the progression of damage that can occur during design basis events, and especially the effects on tube integrity that may result from the dynamic processes associated with depressurization. The staff does not currently have a technically defensible analysis of how steam generator tubes, which may be flawed, will behave under severe accident conditions in which the reactor coolant system remains pressurized. Better databases are needed for the implementation of the condition monitoring of steam generators with 7/8" tubes. A program to detect systematic deviations from the bounding, linear model of flaw growth during operations is needed. The staff needs to develop a more technically defensible treatment of the iodine spiking phenomena associated with design basis events.

CONTENTS

Figures

Table

vii

ABBREVIATIONS

ACRS	Advisory Committee on Reactor Safeguards (NRC)
ANL	Argonne National Laboratory
ATWS	Anticipated transients without scram
AVT	All volatile treatment
$[C]_{ss}$	Steady-state iodine concentration of primary coolant
CFR	*Code of Federal Regulations*
DPO	Differing professional opinion
ECCS	Emergency core cooling system
ECT	Eddy-current techniques
EDO	Executive Director for Operations (NRC)
EOP	Emergency operating procedure
EPRI	Electric Power Research Institute
ERG	Emergency Response Guideline
GDC	General Design Criteria (in Appendix A to 10 CFR Part 50)
GL	Generic Letter
IGA	Intergranular attack
MSLB	Main steamline break
NDE	Nondestructive examination
NEI	Nuclear Energy Institute
NRC	U.S. Nuclear Regulatory Commission
ODSCC	Outside diameter stress corrosion cracking
POD	Probability of detection
PWSCC	Primary water stress corrosion cracking
RHR	Residual heat removal
RWST	Refueling water storage tank
SBO	Station blackout
SCC	Stress corrosion cracking
SF	Spiking factor
SGTR	Steam generator tube rupture

1. INTRODUCTION

This report to the Advisory Committee on Reactor Safeguards (ACRS) of the U.S. Nuclear Regulatory Commission (NRC), concerns technical issues associated with voltage-based alternative repair criteria for steam generator tubes. In a letter dated July 20, 2000 (see Appendix A), the NRC's Executive Director for Operations (EDO) requested that the ACRS examine these issues and provide comments and recommendations for use by the EDO in resolving a differing professional opinion (DPO) concerning the adequacy of protection of public health and safety afforded by the alternative repair criteria documented in Generic Letter (GL) 95-05 **[1]**. The ACRS accepted this request (see Appendix B), and established an Ad Hoc Subcommittee to begin the investigation of the technical issues and prepare a report to the ACRS.

The Ad Hoc Subcommittee was chartered to gather information, analyze relevant facts, and develop draft positions for consideration by the ACRS. Members of the Ad Hoc Subcommittee were ACRS members D.A. Powers (Chairman), M. Bonaca, J. Sieber, and T.S. Kress. Professor R. Ballinger from the Massachusetts Institute of Technology was also a member of the Subcommittee. The Subcommittee was supported by U. Shoop and S. Duraiswamy of the ACRS staff. The Subcommittee also had the benefit of consultants hired by the NRC staff, Professor I. Catton of the University of California at Los Angeles, Mr. J.C. Higgins from Brookhaven National Laboratory, and Dr. R.E. Ricker from the National Institute of Standards and Technology.

The Subcommittee met on October 10, 2000, to discuss the substantial body of literature that has been developed in connection with the technical issues concerning steam generator tube repair. The Subcommittee met with the author of the DPO, Dr. J. Hopenfeld, and Professional Engineer R. Spence, of the NRC staff, on October 11, 2000, to discuss contentions concerning the nature and implementation of alternative repair criteria. On October 12 and 13, 2000, the Subcommittee met with the NRC staff to discuss the staff's responses concerning the contentions identified in the DPO. The Subcommittee examined findings from these meetings on October 14, 2000. This report to the ACRS evolved from those examinations.

An abbreviated discussion of the background concerning steam generators and the need for alternative repair criteria is presented in Chapter 2 of this report. The contentions that have been raised concerning tube repair criteria, technical issues, views of the Ad Hoc Subcommittee, descriptions of the related staff positions, and the data and analyses that are available to support the various positions are presented in Chapter 3 of this report. Significant conclusions and recommendations to the ACRS concerning the issues are presented in Chapter 4.

2. BACKGROUND

Steam generators constitute more than 50% of the surface area of the primary pressure boundary in a pressurized water reactor. This pressure boundary is an important element in the defense in depth against release of radioactive material from the reactor into the environment. Unlike other parts of the reactor pressure boundary, the barrier to fission product release provided by the steam generator tubes is not reinforced by the reactor containment as an additional barrier. Fission products released from the ruptured steam generator tubes can escape into the environment through the secondary side of the steam generator, especially if the release is from a fully pressurized reactor coolant system. Consequently, the integrity of the steam generator tubes must be ensured with high confidence. A rupture of a steam generator tube during power operation of a reactor is an event that must be arrested and mitigated rapidly and with high confidence.

Steam generator tube rupture accidents are considered in the safety evaluation of reactor designs. That is, the pressurized water reactor must be capable of rapid and effective response to the accidental rupture of a steam generator tube. Such accidents are not hypothetical. Worldwide, there have been about 11 steam generator tube rupture events in pressurized water reactors of what might be called the "Western design." Information on these past events is provided in Table 1. The first event listed in Table 1 occurred in 1975 at Point Beach Unit 1. The most recent event occurred in 2000 at Indian Point Unit 2. Clearly, consideration of steam generator tube accidents in the safety evaluation of pressurized water reactors is not anachronistic.

Concern regarding the integrity of the steam generator tubes is not restricted just to the possible rupture of the tubes. Leakage through the tubes from the primary coolant system to the secondary side of the steam generator is also of concern. In the event of a rupture of the main steamline, leakage of reactor coolant through the tubes could contaminate the flow out of the ruptured steamline. In addition, leakage of primary coolant through openings in the steam generator tubes could deplete the inventory of water available for the long-term cooling of the core in the event of an accident. Plant response to a rupture of the main steamline and any leakage of radioactive material through the steam generators is a design basis that must be considered in the safety evaluation of pressurized water reactors.

Table 1

Past Steam Generator Tube Rupture Accidents at Pressurized Water Reactors

Plant	Date	Leak Rate (gpm)	Cause
Point Beach Unit 1	February 26, 1975	125	wastage
Surry Unit 2	September 15, 1976	330	PWSCC in U-bend
Doel Unit 2	June 25, 1979	135	PWSCC in U-bend
Prairie Island 1	October 2, 1979	390	loose parts
Ginna Unit 1	January 25, 1982	760	loose parts and tube wear
Fort Calhoun	May 16, 1984	112	ODSCC at a crevice
North Anna Unit 1	July 15, 1987	637	high cycle fatigue in a U-bend
McGuire Unit 1	March 7, 1989	500	ODSCC in the free span
Mihama Unit 2	February 9, 1991	700	high cycle fatigue
Palo Verde Unit 2	March 14, 1993	240	ODSCC
Indian Point Unit 2	February 15, 2000	150	PWSCC in U-bend

Design basis accident evaluations typically consider only single system failures. Risk analyses consider multiple system failures, including the possibility of human error following accident initiation. Steam generators have been found to be important in determining the risk associated with the operation of pressurized water reactors [2]. Accidents initiated by steam generator tube failure and accompanied by additional failures can lead to melting of the reactor core and massive releases of radioactivity. A substantial fraction of the radioactive material released from the melting core can escape through the ruptured tubes without being subject to the same degree of mitigation by the natural and engineered processes that take place in the reactor containment. Consequently, severe accidents initiated by steam generator tube rupture have been found in some cases to be risk dominant, even though they do not have especially high relative frequencies of occurrence. More recently, there has been concern that severe accidents initiated by other events might place sufficient heat and pressure loads on degraded steam generator tubes that could rupture these tubes. The accident would then progress much like a severe accident initiated by a steam generator tube rupture with large releases of radioactive material to the environment.

The steam generators of interest here are fabricated, using Alloy 600 tubes[1] with carbon steel tube support plates that have drilled holes[2]. Such steam generators have a long history of vulnerability to corrosion. In the earlier days of commercial nuclear power production, the corrosion issue of concern was the uniform erosion of tube wall material, often called "wastage." This uniform erosion was the principal concern when existing requirements were formulated mandating the repair or removal from service of tubes that had lost material to the point that the tube walls had been thinned by more than 40%.

Most of the issues of general corrosion and uniform loss of tube wall material have been solved over the years, through a change in water chemistry, as discussed in Section 3C of this report. Unfortunately, new forms of corrosion have been encountered. Figure 1 presents a schematic diagram of a steam generator, showing the types of corrosion and degradation that have been encountered at various locations in steam generators. A particular issue of concern is stress corrosion cracking of tubes. Such cracking has been observed both on the inside of the tubes (so-called "primary water stress corrosion cracking" or PWSCC) and on the outside of tubes (so-called "outside diameter stress corrosion cracking" or ODSCC). Rather than the uniform erosion of wall material, stress corrosion cracking produces short, narrow cracks that can interlink and penetrate through the wall of the tube. Photographs of stress corrosion cracks are shown in Figure 2. Note that crack width is small, and the crack pattern is not smooth. Cracks can be distinguished as predominantly along the length of a tube (axial cracks) or predominantly around the circumference of a tube (circumferential cracks). Cracks can be present in the "free span" of a tube, the "U-bend" of a tube, or those portions of a tube that are surrounded by tube support plates or the tube sheet. The likelihood of cracking within the region of the tube support plates increases if products of carbon steel corrosion press on the tube (thereby increasing stress), and if the stagnant nature of flow in the narrow, annular regions between the tube and the plate leads to unfavorable chemistry.

In general, stress corrosion cracks are harder to detect than the uniform erosion of material from the tube wall. Nevertheless, techniques based on eddy-currents induced in the tube metal have been developed to identify regions of cracking. These methods have relatively high, but not perfect, reliability for detecting cracks that are larger than about 20% of the wall thickness, even when the cracks are within the regions of the tubes bounded by the tube support plate. The techniques are not nearly so reliable for determining the depth of a crack and, in particular, whether a crack penetrates through 40% of the tube wall thickness. Indeed, cracks of a particular depth can produce quite a range of voltage signals with present eddy-current detection methods. This limitation of the detection

---

[1] Alloy 600 is the generic name. The tradename for the alloy produced by International Nickel is Inconel 600. Carpenter Technologies uses the tradename Pyromet 600 for this alloy.

[2] The issues discussed here are specific to GL 95-05 and Westinghouse steam generators with tube support plates that have drilled holes. The issues do not address other types of generators that have broached hole tube support plates or so-called "egg crate" structures in place of tube support plates, and are less vulnerable to the types of degradation that are discussed here.

methods poses a dilemma. Either some alternative criterion is needed for repairing or removing tubes from service (and such a criterion must be compatible with the capabilities of the detection technologies), or licensees must be required to repair or remove from service any tube that produces any indication of some cracking that could be interpreted as possibly having a depth exceeding 40% of the tube wall thickness.

In Generic Letter 95-05 [1], the NRC staff approved alternative repair criteria that the voltage signal produced by defects in the parts of tubes that are <u>surrounded by tube support plates</u> can be used as the criterion for repairing or replacing the tubes. Tubes with flaw indications producing voltage signals less than 1 volt for 3/4" tubes and less than 2 volts for 7/8" tubes can be left in service. Licensees further committed to use voltage signals from all tube defects to infer the leakage possible from tubes and to predict the condition of the tubes over the next operating cycle of the reactor [3]. In making its proposal, the NRC staff acknowledged that there would be some possibility that cracks of objectionable depth might be overlooked and left in the steam generator for an additional operating cycle. Still, the staff feels that the alternative repair criteria are sufficiently reliable that adequate protection of the public health and safety is still provided.

Dr. J. Hopenfeld of the NRC staff differed with the general view of the staff on the adequate protection afforded the public by the alternative repair criteria. He filed a differing professional opinion on the issue, questioning the adequacy of the technical analyses that support the alternative repair criteria. In particular, he questions:

- analyses of design basis accidents involving steam generator tube rupture and main steamline break
- analyses of steam generator tube rupture during severe accidents
- reliability of eddy-current methods for detecting and sizing flaws in a tube
- predictions of leakage and tube rupture probabilities based on results of eddy-current methods
- estimates of fission product releases that might be associated with design basis accidents

Contentions and technical issues raised by Dr. Hopenfeld, the staff responses to the contentions, and the views of the Ad Hoc Subcommittee are discussed in greater detail in Chapter 3 of this report.

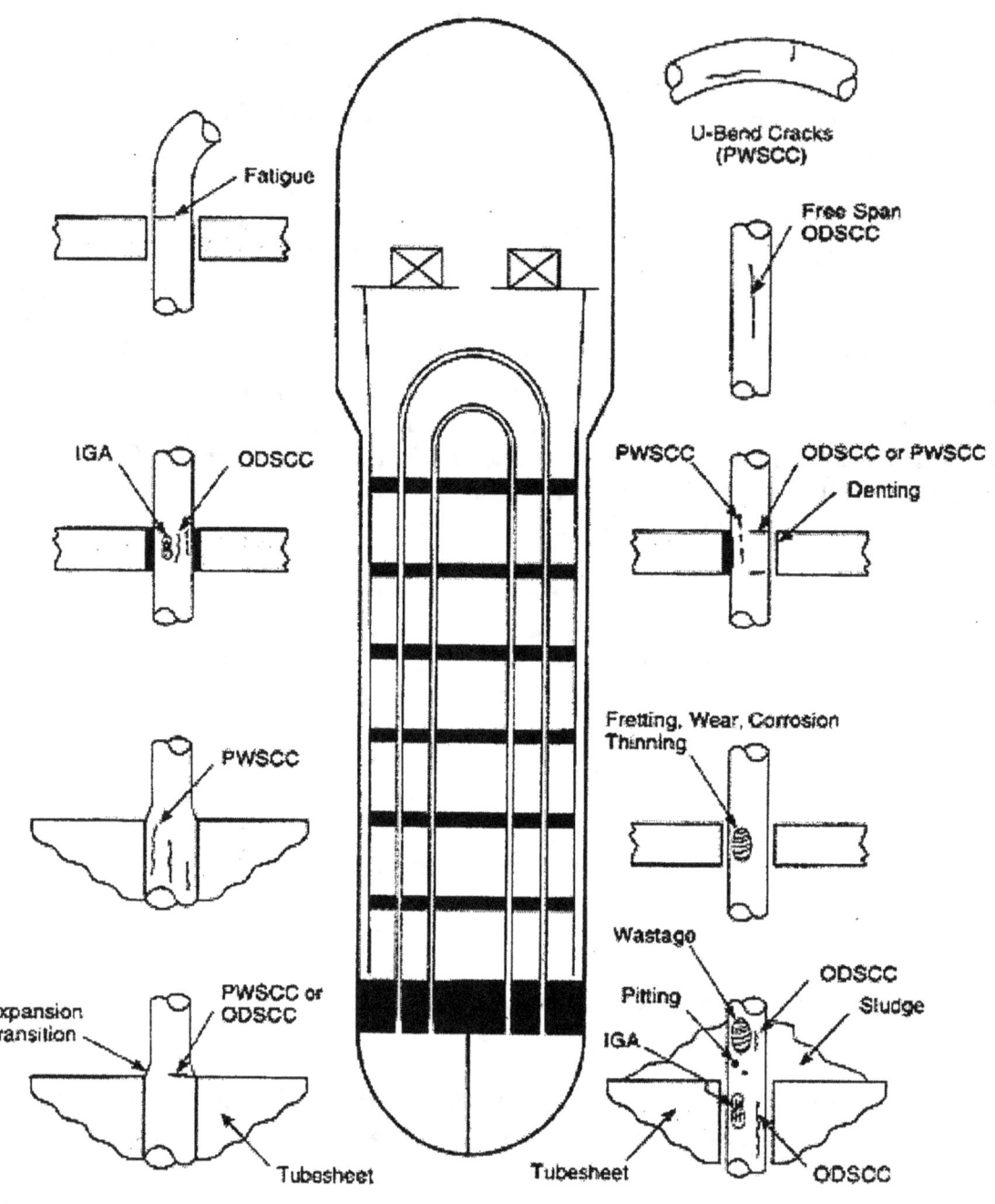

Figure 1. Schematic Diagram of a Steam Generator, and Types of Corrosion and Degradation Observed at Various Locations

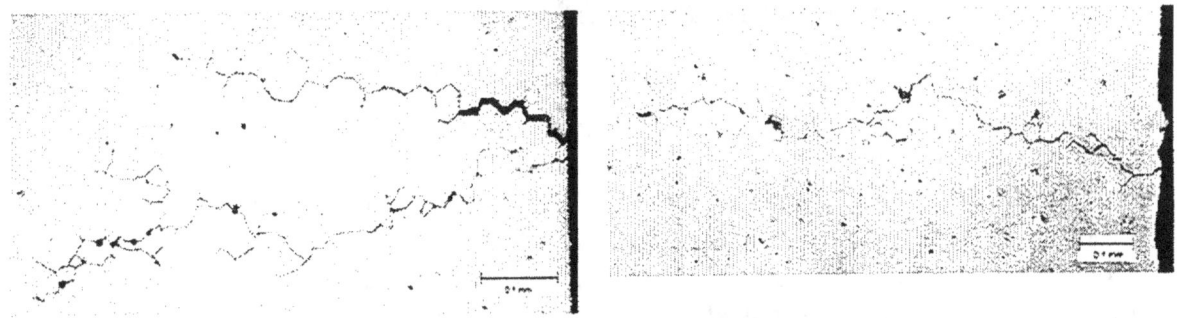

a. Examples of predominantly axial, intergranular stress corrosion cracks. The fiducial marks in these photographs are 0.1 mm long.

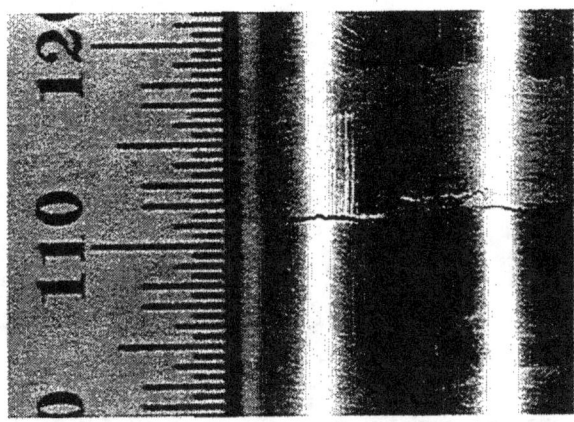

b. Example of a circumferential cracks.

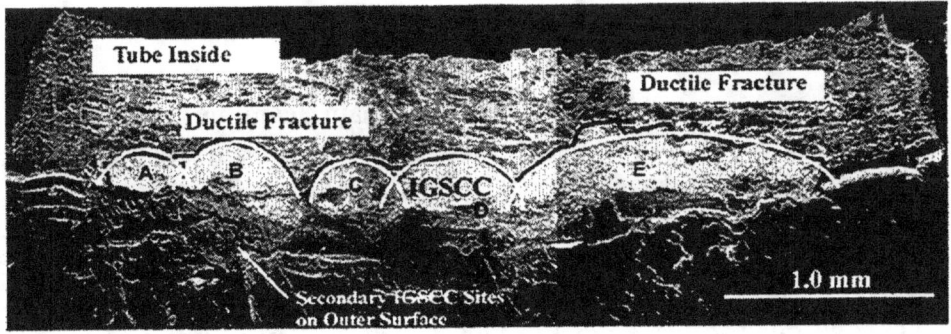

c. Fractograph showing how stress corrosion cracks interlink [12].

Figure 2. Examples of Stress Corrosion Cracks in Alloy 600 Tubes

3. CONTENTIONS AND TECHNICAL ISSUES

A number of contentions were raised by the author of the differing professional opinion (DPO) and the NRC staff concerning the alternative repair criteria. Some of these relate directly to the alternative repair criteria and the prescriptions in Generic Letter 95-05 [1]. Others relate to the analysis of the risk status of a plant, and would arise whether or not the 40% through-wall criterion is replaced by the alternative repair criteria. The contentions are examined in this Chapter. Some effort is made to distinguish those contentions that relate directly to the issue of the repair criteria from those that deal with the risk status of a nuclear power plant.

The topics treated in this Chapter have been grouped into Sections dealing with:

- Reactor accidents
- Human performance
- Stress corrosion cracking
- Nondestructive examination methods and analyses
- Iodine spiking and source term issues

A. Reactor Accidents

Both design basis accidents and severe accidents involving damage to the reactor core need to be considered in connection with the issues of steam generator tube integrity. The design basis accidents are of direct importance to the issues raised in the differing professional opinion. Severe accidents arise in the evaluation of risks associated with any degradation of steam generator tube integrity. Contentions regarding these accidents are discussed in the subsections that follow. In both design basis accidents and severe accidents, human performance has such a significant role that it is treated separately in Section B of this Chapter.

1. Design Basis Accidents

The two design basis accidents of interest here are the main steamline break (MSLB) and the steam generator tube rupture (SGTR). In the past, these accidents have been treated separately. The MSLB was evaluated considering leakage, but not gross flows, from the primary system through the secondary system and out through the break in the steamline. Larger flows, as a result of the rupture of a steam generator tube, from the primary system to the secondary system and out through stuck-open relief valves were treated in the analysis of the SGTR accident. The alternative repair criteria, which are acknowledged to leave cracked and otherwise flawed tubes in service in the steam generator, have reduced the distinction between the two kinds of design basis accidents. The possibility of gross flows from the primary coolant system to the secondary system needs now to be recognized in the analysis of the MSLB accident.

Both the staff and the author of the DPO agree that the alternative repair criteria increase the probability of larger primary-to-secondary flows during the MSLB and SGTR accidents. There are differences in the positions of the staff and the DPO author on the magnitudes of these flows. The staff bases its estimates of the primary-to-secondary leakage during an accident on the analyses discussed in Section D of this Chapter. The staff argues that, in the event of a MSLB accident, the leakage will depend entirely on the number of pre-existing defects in the steam generator tubes, the sizes of these defects, and the pressure difference between the primary and secondary sides of the reactor coolant system. The staff believes that the alternative repair criteria and the condition monitoring program adopted by licensees as a prerequisite for use of the criteria will ensure that any leakages of radioactive material into the environment will be within acceptable limits. The staff expects leakages will be less than 100 gallons per minute (gpm), usually.

The author of the DPO does not have confidence in the methods used by the staff to estimate leakage. Specific concerns he has raised about these methods are discussed in Section D. The more significant contention made by the author of the DPO is that the staff has neglected important phenomena and processes that will damage the tubes and produce leakage in ways that are not recognized by the analytic methods that have been approved by the staff. The phenomena that could lead to damage of multiple tubes cited by the DPO author include:

- shock waves and severe, sympathetic vibrations during depressurization of the reactor coolant system
- jet cutting of adjacent tubes by high pressure fluids laden with particulate emerging from cracked tubes
- movement of the tube support plate and other steam generator internals by the blowdown forces
- growth of pre-existing flaws and opening of cracks that are plugged by corrosion products

The DPO author contends that the flow from the primary side to the secondary side of the reactor coolant system could be more than an order of magnitude greater than what can be estimated using the methods approved by the staff. The DPO author argues that phenomena omitted from the staff's analyses cause the larger primary-to-secondary flows. In the view presented in the DPO, these flows can be large enough that reactor operators may not have the time or the ability to recover long-term cooling of the reactor core.

a. Damage Progression

A key thrust of the arguments made in the DPO is that processes not considered by the staff can lead to greater leakage from the primary side to secondary side of the reactor coolant system. The leakage induced by these processes can be sufficiently large that it substantially reduces the time available for the operators to throttle flow from the emergency core cooling system (ECCS) and avoid depletion of the coolant inventory. Maintenance of this inventory in the refueling water storage tank (RWST) makes possible the long-term cooling of the reactor core by continued ECCS injection until the residual heat removal (RHR) cooling mode can be entered. The phenomena and processes the staff did not consider in its estimation of leakage are discussed in the subsections that follow.

b. Blowdown Forces

> **Contention:** **Depressurization of the reactor coolant system during a main steamline break will produce shock waves and violent, sympathetic vibrations that will cause cracks to form, to grow and to unplug, leading to much higher leakage from the primary-to-secondary sides of the reactor coolant system than has been considered by the NRC staff.**

There has not been a MSLB accident in an operating reactor to date, but there have been at least two occasions of analogous blowdowns in reactor systems during pre-operational testing. Professional Engineer R. Spence of the NRC staff provided the Ad Hoc Subcommittee with a graphic description of the violence of these depressurization events. The accounts noted sonic booms and violent vibrations of the reactor coolant system and support structures. These descriptions made plausible the arguments that the blowdown forces could induce additional damage to the tubes in the steam generator and, consequently, there could be much larger primary-to-secondary flow of coolant. The accounts of these events also made it plausible that operators would experience considerable distraction during such an event. Such distraction is not well reproduced in simulator exercises dealing with a MSLB accident.

The staff does not contest this contention concerning additional phenomena that could further damage the steam generator tubes. The staff has initiated an investigation of this issue that could result in the establishment of a generic safety issue.

The Ad Hoc Subcommittee also finds that this contention of the DPO has merit and deserves investigation. The issue, however, affects any consideration of steam generator tube integrity, regardless of the criteria adopted for tube repair or removal from service. This issue, at the current level of understanding, cannot be used to judge the adequacy of the alternative repair criteria described in GL 95-05. The Subcommittee notes that thermal-hydraulic codes usually employed by the staff for safety analyses are poorly suited to address the issues raised by this contention. The Subcommittee urges that investigation of this issue be completed expeditiously.

c. Jet Cutting of Adjacent Tubes

> **Contention:** **High pressure, particulate-laden fluids flowing from a cracked steam generator tube can pierce adjacent tubes.**

It is known that very high pressure gas and liquid flows, especially with entrained water droplets or particles, can readily cut through structural metals. Conventional uses of this machining method employ pressures about an order of magnitude higher than the pressures expected during a reactor accident. To date, this hypothesized mechanism for the progression of tube damage and enhanced leakage has not been observed in SGTR accidents.

The staff has undertaken experiments and computational fluid dynamics analyses to examine the possibility that flows from one cracked steam generator tube could erode adjacent tubes. Though the studies are not yet complete, results to date have been negative even in cases in which particles were entrained in the high pressure flow. The computational fluid dynamics analyses suggest flows do not lead to extensive, high velocity impacts of entrained particles on the surfaces of adjacent tubes. The analytic predictions appear to be consistent with observations from experiments done to date that indicate little more than burnishing of surfaces exposed to flows of the type expected in reactor accidents.

The Ad Hoc Subcommittee concluded that damage progression by high pressure fluids penetrating adjacent tubes is of low enough probability that it can be neglected in accident analyses. This judgment needs to be confirmed by completion of the planned experiments and analyses. It is particularly important that tests are conducted for sufficiently long periods to define rates of material erosion. Such rate data are needed to ensure that jet erosion or cutting will not occur over a protracted timespan of hypothesized accidents. Greater confidence in a negative result from these studies will be gained if the experiments and analyses are extended to define the conditions where fluid flows will penetrate adjacent tubes, even if these conditions are well beyond anything expected in a reactor accident.

d. Crack Unplugging

> **Contention:** **Forces involved in the blowdown and leakage can cause cracks plugged with corrosion products to leak. Corrosion products in the annular gap between the tubes and holes in the tube support plate can be expelled, allowing otherwise occluded cracks to leak.**

The DPO author contends that the procedures used by the staff to predict leakage do not account for the blowdown and other forces that could cause cracks that are plugged with corrosion products to open and contribute to the leakage. The staff argues that the cracks that have not already been considered to contribute to the leakage are typically quite narrow ("tight" cracks), and are not plugged with corrosion products in a way that would prevent detection of leakage in tests used to formulate either the probability of leakage database or the correlation between leak rate and the voltage signal produced by the crack. The Ad Hoc Subcommittee has found the staff's position to be persuasive.

The DPO author further contends that forces during blowdown and the force of fluids escaping through cracks could expel corrosion products that occlude cracks in tubes within the tube support plates. The staff has responded to this contention with reference to work done in France on the structural capabilities of "crud" and corrosion products in holes and on the tube support plate. The staff also notes that leakage tests done on tube flaws from within the tube support plate were done without the plate present. Consequently, any inhibition provided by the corrosion products within the steam generator would not be reflected in the leakage estimation procedure. The Ad Hoc Subcommittee has found the staff's position persuasive on this issue.

e. Tube Support Plate Lift

> **Contention:** **Tube support plates can be lifted during the sudden depressurization of a main steamline break and this can cause cracks in tubes to penetrate through the tube walls and lead to additional flow from the primary coolant system to the secondary side of the coolant system.**

This seems to be a plausible contention, and the staff has not produced analyses or test results to refute it. The staff seems to view plate movement as predominately axial, and did not provide analysis of lateral movements or bending motions that might damage steam generator tubes. Analysis of movement of the tube support plate during depressurization is a difficult undertaking, even without considering the dynamic effects and system vibrations described by Professional Engineer R. Spence. It appears that the usual computer codes (such as RELAP5) have not been qualified adequately for this analysis in the view of the staff. Use of such one dimensional models for analysis of multidimensional situations must be done with care and verified by comparison to applicable data. High-quality experimental data that could be used to qualify the codes do not seem to be available. The frequently cited MB-2 tests do not appear suitable for such code qualification because of improper scaling of the test volume.

2. Severe Accidents

Two classes of severe reactor accidents need to be considered in assessing the risks associated with maintenance of steam generator tube integrity. The first class involves accidents initiated by steam generator tube rupture and accompanied by failures of additional systems or human errors. This class of accidents has been considered in probabilistic risk assessments, including those done by the staff [2] and by licensees [4]. This class of accidents has the peculiar feature of being risk dominant, although the accidents are not frequency dominant. That is, the consequences of a SGTR accident progressing to core damage are severe, even though the expected frequencies are low. Severity of the accident consequences arises, of course, because gases laden with radioactive aerosol pass from the primary coolant system through the secondary system and out a stuck-open safety relief valve without passing into the reactor containment. Consequently, natural and engineering processes within the containment do not have the opportunity to mitigate the potential release of radioactive material to the environment in such a "containment bypass accident."

The issues of leakage during the design basis MSLB and SGTR accidents discussed above, and the reliability of operator actions discussed in Section B of this Chapter, treat the pertinent issues of severe accidents initiated by steam generator tube ruptures. The Ad Hoc Subcommittee notes, however, that analyses of the consequences of these severe accidents done in the past [2] did not model well the mitigation of the radionuclide release that could occur in the secondary side of the plant, and may have overestimated the consequences of these accidents. The ARTIST program planned in Switzerland to study experimentally the mitigation that could occur during radionuclide transport through the secondary side of a steam generator may provide data needed to model better the natural mitigation of the source term associated with steam generator tube rupture severe accidents.

The second class of severe accidents of interest can be initiated by something other than a steam generator tube rupture, and the primary system remains pressurized. Important examples of such accidents are the station blackout (SBO) accidents, very small break accidents, and anticipated transients without scram (ATWS). These severe accidents are of interest because they impose high heat and pressure loads on the steam generator tubes. These loads imposed on sufficiently degraded tubes could cause gross tube failure. The accident would evolve to become much like an accident initiated by a steam generator tube rupture, and released radioactivity could bypass the reactor containment. Such evolution of the accidents could result in substantially greater severe accident consequences.

Contention: **Severe accident sequences in which the primary system remains pressurized are more likely to evolve into steam generator tube rupture accidents than the staff predicts.**

The staff has analyzed the possible evolution of severe accidents and the loads that they place on the steam generator tubes using computer models like SCDAP-RELAP5. These analyses [5] are predicated on the assumption that the "loop-seals" in the primary reactor coolant system remain in place, so there is not a gross convection through the reactor coolant system[3]. Instead, a countercurrent natural circulation of steam from the degrading reactor core through the steam generator tubes and back to the core is established. A schematic diagram of this natural circulation flow is shown in Figure 3. The analyses done by the staff indicate that heat is transferred to other parts of the primary reactor coolant system by the natural circulation flow, and failure at other locations is likely to occur before a steam generator tube undergoes gross failure as a result of creep rupture. Typically, the surge line is predicted to fail first, followed by failure of a reactor nozzle. The analyses also show that the timing of the failures at the various locations are not greatly different.

---

[3] If a loop seal is opened and the coolant level in the pressure vessel is sufficiently low, an overall natural circulation of high-temperature steam through the steam generator and back into the vessel develops (see the left side of Figure 3.) This overall natural circulation is thought to impose sufficiently high heat loads on the steam generator tubes that the tubes fail whether they are flawed or not. Detailed analyses have not been done, but it is known that additional complexities can develop in the analyses of such scenarios, including episodic re-establishment of the loop seals.

The codes used by the staff for these analyses are lumped parameter codes and do not make reliable predictions of natural circulation and other phenomena involving momentum. The codes are "tuned" by comparison to experimental results, such as the results obtained with a one-seventh scale model of a pressurized water reactor coolant system [5]. These tests have been criticized because they did not involve proper dimensional scaling of the steam generator portion of the test fixture. One consequence of this imperfect scaling is that the mixing of the hot fluid entering the plenum with cold fluid coming out of the steam generator may have been overestimated. This mixing significantly reduces the heat loads imposed on the steam generator tubes and the delay in the creep rupture of the tubes. Sensitivity analyses that have been done as part of the analyses have not varied parameters over the entire plausible range, and have not examined simultaneous variations of multiple parameters [5].

Another concern is that the tests did not simulate the leakage through the steam generator tubes that might occur. Sufficiently large leakage could bias the flow of hot fluid entering the lower plenum below the steam generator tube support plate so that there would be much less mixing and much less cooling of this fluid. Again, sensitivity studies of the computed results have not explored the plausible range of parameters.

The Ad Hoc Subcommittee concluded that the issue of the possible evolution of severe accidents to involve gross failure of steam generator tubes and bypass of the containment is not yet resolved. The Subcommittee also believes that the issue needs consideration regardless of the criteria adopted for the repair and replacement of steam generator tubes. The issue may be affected by any enhanced leakage that comes from the use of the alternative repair criteria described in GL 95-05. But, the issue is just one of several significant uncertainties that now exist in the assessment of severe accident progression.

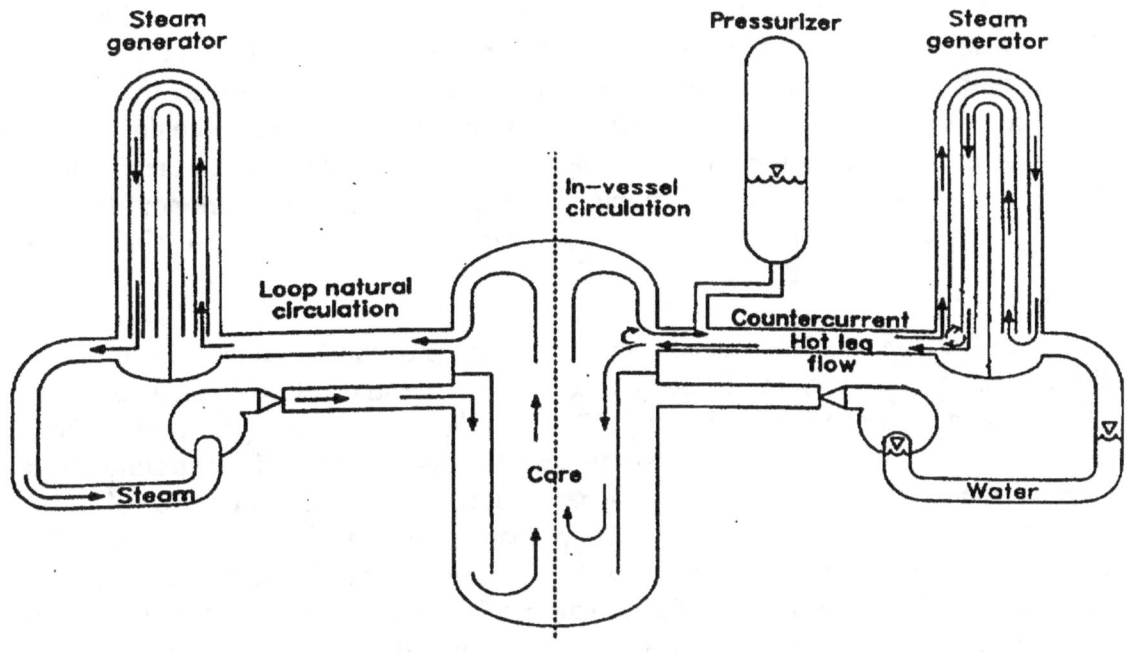

Figure 3. Schematic Diagram of Natural Circulation in the Primary Reactor Coolant System. On the left, full loop natural circulation is shown. On the right, countercurrent natural circulation through the steam generator is shown.

B. Human Performance

The staff and the author of the DPO agree that there are key operator actions that must take place during MSLB and SGTR accidents. During such accidents, coolant is being expelled from the primary side to the secondary side, and out through a break or a stuck-open relief valve in the secondary side. The operators must take action to mitigate the release of contaminated coolant and preserve water inventory in the reactor coolant system and the RWST for as long as is needed to depressurize and to cool the primary side sufficiently to activate the RHR system. The key actions of the operators are typically reducing the primary coolant system pressure to reduce leakage flow to the secondary side, cooling of the primary reactor coolant system by feeding the secondary side of the intact steam generators, and throttling of the ECCS to preserve the water inventory if leakage through the steam generator tube rupture is large. These activities are not all simple, and it is generally agreed that a steam generator tube rupture is one of the more challenging events for operators.

Where the staff and the DPO author disagree is on the probability that the operators will accomplish these tasks successfully in all likely cases. The essential role of human performance in the disagreement between the staff and the DPO author is illustrated by a simple calculation of the probability of containment bypass for an accident involving a main steamline break. The staff predicts the frequency of containment bypass induced by a main steamline break to be 7×10^{-7} yr^{-1}. This estimate is based on an operator error probability of 1×10^{-3}. The author of the DPO takes the probability of a MSLB accident to be 1×10^{-4} yr^{-1}. He argues that the flow from the primary to the secondary coolant system and out through the break, will be so large that the operators will not have time to prevent depletion of the coolant inventory and that the operators will be considerably distracted by the dynamic events of the depressurization associated with the main steamline break. He estimates the operator error probability to be essentially 1.0, so that the probability of containment bypass is also 1×10^{-4} yr^{-1}. The author of the DPO equates containment bypass to a large early release of radioactivity. In analogous calculations for a SGTR accident with an initiating probability of 1×10^{-3} yr^{-1}, the DPO author argues that because of damage progression leading to large flows from the primary to secondary systems, the operator error probability is about 0.1. Again, he concludes that, for the steam generator tube rupture sequence, the probability of containment bypass is about 1×10^{-4} yr^{-1}. This can be compared to the staff's estimate of 5×10^{-7} yr^{-1} for containment bypass due to a steam generator tube rupture and a stuck-open relief valve. Again, in developing this estimate, the staff estimates that the probability of operator error is 10^{-3}.

Contention: **The very low probability, 10^{-3}, that the operators will fail to perform tasks needed to establish long-term cooling of the core is overly optimistic.**

Factors that need to be considered in estimating the reliability of human performance include:

- availability of readily understood procedures to carry out the needed operations

- operator experience and training in these procedures
- ambiguity in the causes of events confronting the operators, including difficulties with the human-system interface in the control room
- communications within the operations staff
- stress on and distraction of the operators during the time that the plant condition must be diagnosed and the procedures must be carried out
- the time available for the operators to diagnose the situation, identify the correct procedures, and carry out the appropriate plant evolutions

The staff and the DPO author agree that appropriate, symptom-oriented procedures for responding to MSLB events exist at all plants. These procedures for the Surry plant are described in NUREG-1477 [6]. Specifically, the operator is to follow Emergency Response Guideline (ERG) E-0, "Reactor Trip or Safety Injection," until he observes that pressure in the leaking steam generator is decreasing in an uncontrollable way. Then, the operator is directed to enter ERG E-2, "Faulted Steam Generator Isolation," to identify and isolate the faulted steam generator. ERG E-2 requires that the operator perform a secondary radiation check to identify leaking or ruptured steam generator tubes. While the check is being performed, the operator enters ERG E-1, "Loss of Reactor or Secondary Coolant," trips the reactor coolant pumps and performs a check on whether safety injection flow should be reduced. Once the check verifies that a steam generator tube rupture has occurred, the operator enters ERG E-3, "Steam Generator Tube Rupture." ECA-3.1, "Steam Generator Tube Rupture with Loss of Reactor Coolant — Subcooled Recovery Desired," is used for a steam generator tube rupture in combination with a main steamline break. This procedure contains instructions for limiting the leakage of reactor coolant for multiple steam generator tube ruptures, and is designed to respond to a class of accidents in which leakage from the reactor coolant system cannot be stopped until cold shutdown is achieved. Key recovery objectives of this procedure are to maintain sufficient inventory in the reactor coolant system to ensure core cooling, and to minimize leakage of the inventory to conserve makeup water and to limit radiological releases. If the operator identifies a low inventory in the RWST, combined with containment sump water less than expected, the operator enters ECA-3.2, "Steam Generator Tube Rupture with Loss of Reactor Coolant — Saturated Recovery Desired," to add borated water to the RWST and to allow a saturated reactor coolant system condition, which is intended to minimize further leakage from the reactor coolant system.

It is important to recognize that the symptom-oriented emergency operating procedures are designed so that the operators will depressurize and cool the primary system to enter the RHR cooling mode even if they fail to identify the steam generator tube leakage. If the initiating event is a steamline break and consequential steam generator tube rupture, the reactor coolant system will depressurize and cool without operator action. The depressurization and cooling rates increase with the magnitude of the flow through the ruptured steam generator.

The staff and the author of the DPO agree that operators are trained in these procedures. They disagree on the effectiveness of the training. The DPO author notes that in both simulator exercises and in actual SGTR events that did not involve a steamline break, there have been delays in operator diagnosis and response to events. Though needed plant evolutions have been carried out successfully, these operations were not always carried out within the 30-minute time window expected in the emergency operating procedures. The DPO author further notes that in tests at the Halden simulator, aspects of the performance of several crews were rated as "poor."

The staff has responded to these contentions by noting that the evolutions do get accomplished successfully. The 30-minute time window is a regulatory requirement that applies to the SGTR event alone, and is based on the need to depressurize the reactor coolant system and limit releases from the secondary side within the time assumed in the design basis accident analysis. The 30-minute time window is not a requirement to prevent core damage and extensive release of radioactivity from the plant.

In the absence of a steamline break, depressurization of the primary system to stop leakage from the steam generator in 30 minutes can be a challenging task. Mean times for operator actions in 10 actual SGTR events are provided below [7,8].

Operator Action	Mean Time
Diagnosis of steam generator tube rupture	16 minutes
Depressurization	31 minutes
Throttling of emergency core cooling system	14 minutes
Total time	61 minutes

In none of these events was there core damage, and the associated releases of radioactive materials from the plants were small. Human error probabilities from the *Handbook of Human Reliability Analysis with Emphasis on Nuclear Power Plant Applications* [9] indicate that the probability of failure of control room personnel to diagnose an abnormal event within one hour is 1×10^{-3}. The probability that the operators will fail to diagnose an abnormal event decreases with time, but only slowly. When the time available is 24 hours, the failure probability is estimated to be 1×10^{-4}. Human error probabilities from seven individual plant examinations for Westinghouse pressurized water reactors for "early isolation of the ruptured steam generator and stabilization of the reactor coolant system and steam generator pressure prior to steam generator overfill" varied from 1.1×10^{-3} to 5.0×10^{-2}. For the action, "depressurize the primary below that of the secondary with high pressure injection available," the human error probabilities varied from 9.3×10^{-4} to 5.0×10^{-2}.

In the event of a steamline break with rupture of a steam generator tube, the criterion for success is not isolation of the steam generator in 30 minutes. It is the successful depressurization and cooling of the primary system to allow entry into the RHR cooling mode before the water inventory is depleted. These actions are discussed in the Westinghouse Emergency Response Guidelines. With just one failed steam generator tube during a main steamline break, the operator actions are diagnosis, depressurization and cooldown, throttling the emergency core cooling injection, and initiation of the RHR system. The time needed to accomplish these actions and prevent core damage is estimated to be 2.9 hours. The time available for these actions varies from 13.1 to 4.3 hours. The probability of not recovering from the event is estimated to vary from 9.6×10^{-4} to 0.19 depending, of course, on how much more time is available to do the job than the job requires.

The DPO author has noted that the experience gained through simulator training and in actual SGTR events may not be reliable indicators of operator performance during MSLB events. The dynamic nature of the protracted depressurization of the secondary system accompanied by sonic booms and violent vibrations of the reactor coolant system and support structures may distract operators considerably. No information is available on the level of distraction that might occur, since there has not been a MSLB event at an operating plant and the distraction conditions cannot be reproduced well in simulator exercises. Professional Engineer R. Spence has indicated, based on personal experience during a depressurization event at an unfueled reactor in the early 1970s, that even oral communications within the control room can be challenging.

The Ad Hoc Subcommittee concluded that effective, high-quality procedures for the response to MSLB and SGTR accidents are available. Operators are trained in, and can follow, these procedures to depressurize and cool the primary system for entry into the RHR cooling mode. The crucial issue is the time available for the operator actions, that is discussed later in this report. This issue is directly related to the contentions concerning the alternative repair criteria and the magnitude of flow from the primary side to the secondary side of the reactor coolant system.

The Ad Hoc Subcommittee also concluded that the issue of operator distraction and stress caused by the poorly understood dynamic features of depressurization during a MSLB accident is important. This issue transcends the issues of the alternative repair criteria, and arises regardless of the repair criteria used for steam generator tubes. This issue of operator distraction needs to be borne in mind as the staff investigates the generic safety implication of the dynamic depressurization processes.

The essential debate between the author of the DPO and the staff revolves around the time available for the operators to respond to an event. Indeed, the time available relative to the time required to do a task is a very important factor in determining human error probabilities. The staff has performed analyses of the event timing considering flows up to 1000 gallons per minute from the primary side to the secondary side of the reactor coolant system. These analyses are reported in NUREG-1477 [6]. Assuming no operator action to depressurize and cool down the primary reactor coolant system, the water inventory will be depleted over 8 hours to the point that insufficient water will remain to complete depressurization and cooldown for entry into the RHR cooling mode and support the long-term cooling of the core. If the operator depressurizes the reactor coolant system to reduce leakage to the secondary side of the system, about 20 hours will be available before cooldown must begin to avoid depleting the RWST inventory. Based on this time interval of 8 to 20 hours, training, and the availability of procedures, the staff has concluded that the probability of failure of the operator to respond effectively to the event is about 10^{-3}. The Subcommittee concluded that this estimate is in accordance with the current state-of-the-art for estimating human error probabilities.

On the other hand, the damage progression hypothesized in the DPO could lead to greater flows from the primary side to the secondary side of the reactor coolant system. The staff has sponsored analyses of operator actions for up to 15 tube ruptures [8]. Higher primary-to-secondary flows can shorten the time available for operator actions considerably, but, at the same time, these higher flows make it easier for the operator to diagnose and respond to the accident. For example, assuming three failed steam generator tubes, the operators will have 8 hours to RWST depletion if they throttle the emergency core cooling injection at 1 hour from the beginning of the accident. If they fail to throttle the injection until 3 hours after the start of the event, the RWST inventory will be depleted in 5 hours. If 10 tubes fail, water inventory will be depleted in 3.5 hours if the emergency core cooling injection is throttled at 1 hour after the start of the event. The probability that the operators will fail to recover in this case is estimated to be between 3×10^{-3} and 2×10^{-2}. Failure of 15 tubes requires that the operators throttle the emergency core cooling injection quickly if water inventory is to be preserved. The time required to perform this action is estimated to be 40 minutes, and the time available to complete the action is 66 minutes. The probability that the operators will fail to perform these actions is estimated to be 0.2. The study shows that accident processes can affect the necessary operator actions in other ways. For instance, the operator must depressurize the reactor coolant system to enter RHR when one to three steam generator tubes are ruptured. When 4 to 10 tubes are ruptured, the system depressurizes itself and cannot repressurize sufficiently to require additional operator actions to depressurize the reactor coolant system. This natural depressurization reduces the rate of leakage from the primary system to the secondary system and effectively eliminates the concern about operator actions to depressurize the system. Still, the operators must act to throttle the safety injection to limit the loss of coolant inventory outside the reactor containment.

The Ad Hoc Subcommittee concluded that shortened times available to the operator when more extensive tube ruptures take place do reduce the reliability of operator actions to preserve long-term cooling capabilities. The failure probabilities can rise from 10^{-3} to ~ 1.0, depending on the number of failed steam generator tubes. The empirical evidence from steam generator tube rupture events that have occurred to date suggests that the likelihood of an event involving multiple tube ruptures is much less than the likelihood of an event involving a single tube rupture. There is, however, no such empirical database to support arguments concerning the likelihood of multiple steam generator tube ruptures during main steamline ruptures. Consequently, scenarios involving higher operator failure probabilities should be included in assessing risk changes associated with the alternative repair criteria. Risk evaluations should also include examination of the mechanisms for damage progression, which has not been observed in steam generator tube rupture accidents to date, but may occur as a result of dynamic processes during main steamline break depressurizations of the reactor coolant system. The effects of the dynamic events on operator performance both with respect to the time available for required responses and the level of operator distraction need to be evaluated.

C. Stress Corrosion Cracking

Licensees have committed to comply with a condition monitoring program that has to be considered a part of the alternative repair criteria [3]. This condition monitoring program involves both an assessment of the degradation status of a steam generator at the end of an operating cycle and a prediction of the progression of the degradation over the next operating cycle. The assessment and predictions have been the sources of many number of contentions in the DPO, both with respect to the methods (the feasibility of predicting crack growth) and the databases (utility of laboratory data).

To discuss these contentions effectively, it is useful to provide some background regarding the nature of steam generators and the nature of the stress corrosion cracking that afflicts these generators. This background is provided in the next subsection of this Chapter. Contentions regarding the state of knowledge concerning metallurgy and stress corrosion cracking are discussed following the background subsection. Contentions dealing with the analysis and interpretation of results obtained in the nondestructive examination of steam generators and the condition monitoring program are also discussed in Section D of this Chapter.

1. Steam Generator History

A schematic diagram of a steam generator is shown in Figure 1. Steam generators of interest here are made with Alloy 600 (75 weight % nickel, 15 weight % chromium, and 10 weight % iron)[4] tubes. Alloy 600 was selected as the tube material after it was found that stainless steel alloys suffered stress corrosion cracking due to chloride in-leakage on the secondary side. This change in tube material eliminated the chloride cracking problem. Unfortunately, Alloy 600 has proven to be susceptible to stress corrosion cracking in the purer waters typical today of the primary and secondary sides of pressurized water reactor coolant systems.

Typically, the tubes have been "mill annealed." Mill annealing involves a final heat treatment of the unbent tubes at 925 to 1050°C and, then, allowing the tubes to cool in a manner that is largely uncontrolled. This heat treatment as well as other processing conditions have lead to a great deal of variability in the detailed microstructure of the tube alloy and, consequently, variability in the susceptibility of the tubes to corrosion processes. In some of the later steam generator models, an additional thermal annealing of the tubes has been done, which appears to delay the onset of corrosion of the type of interest here. Following the annealing, tubes were straightened and, in some cases, subjected to grinding. The straightening process left residual stresses in both the internal and external surfaces of the tubes. Any grinding operations left residual stresses in the outer surfaces of the tubes.

Other pertinent parts of the steam generators are the tube support plates and the tube sheet. These parts are fabricated from carbon steel with drilled holes. There can be a substantial galvanic potential between the steam generator tubes and the carbon steel structures. The rationale for the

---

[4] Replacement steam generators being installed in many plants are made with Inconel 690 alloy tubes and stainless steel tube support plates with broached holes. These new generators are thought to be much less susceptible to the corrosion problems discussed in this and the next subsection.

remarkable choice of easily corroded carbon steel for the tube support plates and the tube sheet is obscure.

Final fabrication of the steam generators involved bending the tubes and inserting them in the tube sheet. The bending operation leaves substantial stresses in the bent portion of the tubing. These stresses are, of course, highest in the tubes with the smallest bend radius. In early designs, the tubes were inserted into the tube sheet and partially expanded to form leak-tight joints, and the tubes were welded to the sheet on the primary coolant system side. Residual stress left by the partial expansion depends on the method used to do the expansion. Perhaps of more importance, the partial expansion left a long annular space between the tube and the tube sheet. In this annular space (often called a "crevice"), aqueous chemistry substantially different than the bulk coolant chemistry could develop, and corrosion processes different than in the free spans of the tubes could occur. Newer steam generators eliminate this crevice by expanding the tube along the full length of the tube sheet. Some licensees have modified existing steam generators to have full expansions of the tubes in the tube sheet. Different aqueous chemistry can develop also in the annular spaces where tubes pass through drilled tube support plates. Deposits of corrosion products that form in these annular spaces can grow to the point that they press on the tubes and create additional stresses.

The chemical conditions maintained in the steam generators to control corrosion have evolved over the years [10]. The original water chemistry control scheme for steam generators was based on the experiences in the fossil energy industry. The focus was on the control of the acid concentration (pH) and the oxygen concentration to minimize the general corrosion of the carbon steel components of the steam generators. Chemistry control was achieved by using a combination of phosphate for pH control and sulfite for oxygen control. Hydrazine was also used for oxygen control during low-temperature operations. Reactions of the additives with contaminants in the water created a soft "sludge" that could be removed from the steam generator by blowdown. The disadvantages of this system of chemistry control included the production of dissolved solids, the formation of large quantities of sludge that collected on the tube sheet and the tube support plates, and the potential for the formation of high concentrations of hydroxide ions in the water. The formation of hydroxide along with the complexing nature of phosphate led to the wastage of tube materials. The collection of sludge on the tube sheet aggravated the unfavorable local chemistry and, again led to wastage of the tubes.

For these reasons, the industry has shifted to ammonia or some volatile amine for the control of pH and hydrazine for oxygen control (the so-called "all volatile treatment" or AVT chemistry control). This change has eliminated problems that arise from formation of dissolved solids. There has been, however, a loss in the buffering capacity of the system and a greater sensitivity to inleakage of contaminates. The sludge formed with the all volatile chemistry control system tends to be harder and less easily removed from the system. Also, there is now much greater control of the purity of the water used in steam generators. In the past, control of contaminants to the level of parts per million was common. Now, control of contaminants to the level of parts per billion is sought.

2. Occurrence of Stress Corrosion Cracking

The various types of corrosion that have been observed in steam generators over the years are shown in Figure 1. As noted above, the corrosion of interest in the early days of steam generator usage was wastage occurring uniformly along some length of a tube. Of primary interest now are the instances of stress corrosion cracking on both the inside and outside of tubes, and a progenitor of outside diameter stress corrosion cracking called intergranular attack (IGA). Stress corrosion cracking requires three conditions to exist simultaneously:

- a corrosive environment locally
- a tensile stress greater than some threshold value
- a material with a microstructure susceptible to this type of cracking

The evolution in the various forms of stress corrosion cracking observed in steam generators has been driven more by the evolution of operational stresses on the tubes and changes in the coolant chemistry than it has by any change in material susceptibility. The shift to the all volatile chemistry control mentioned above reduced substantially the buffering capacity that was available from the phosphate-based chemistry control. This results in the concentration of contaminants and hydroxide ions in solutions in the crevices in the tube sheet and in the tube support plates. In the case of tube support plate crevices, concentration factors of 2×10^4 have been measured [11]. The concentration of contaminants in the crevice regions accelerates the corrosion of the carbon steel components. The products of this corrosion produce stresses on the tubes (the so-called "denting" problem). Stress corrosion cracking observed on the inside and outside of tubes in the regions of the tube support plates derive, in large part, from the stresses produced by these corrosion products. Stress corrosion cracking on the inside of tubes (primary water stress corrosion cracking) occurs in highly stressed regions such as roll transitions, tube expansions in the tube sheet, and in the region of the U-bend. Denting can also aggravate the stresses in the U-bend region by displacing the tube. Stress corrosion cracking on the outside of tubes (outside diameter stress corrosion cracking) is commonly observed in the crevice regions of the tube support plates and in the tube sheet, as well as in sludge piles that accumulate especially on the tube sheet.

3. Crack Initiation and Growth

Tube perforation by stress corrosion cracking occurs in two stages — crack initiation and crack growth. The first stage, crack initiation, occurs after an incubation time, which is a function of the material susceptibility, temperature, and stress. Laboratory testing shows that the cracks are affected by the local tensile stress at lengths as short as a few grain diameters [12]. The dependence of the time for crack initiation on stress and temperature is of the form:

$$\frac{1}{t_i} \propto M\sigma^4 e^{\frac{-Q}{RT}}$$

where:

t_i = initiation time

M = material parameter

σ = total tensile stress

Q = activation energy

R = gas constant, and

T = absolute temperature

The material parameter in the above equation is related to the degree to which grain boundaries are covered with carbide, among other things. The activation energy is usually found to be in the range of 45—70 kcal/mole, so crack initiation is quite sensitive to the operating temperature.

The second stage is the growth of the initiated crack, and this growth depends on the stress intensity at the crack tip as well as temperature:

$$\frac{da}{dt} = A \exp\left(\frac{Q_g}{RT}\right) (K_I - K_{th})^B$$

where:

a = crack length

A = pre-exponential factor

Q_g = activation energy for crack growth ~ 25 kcal/mole

K_I = stress intensity factor at the crack tip (MPa$\sqrt{(m)}$)

K_{th} = threshold value of the stress intensity factor

B = parameter on the order of 1.5

R = gas constant

T = absolute temperature

It should be noted that the stress intensity factor in this equation depends on the crack length.

The availability of these functional forms for the processes of crack initiation and growth has made possible the study of stress corrosion cracking under conditions that accelerate the process relative to what it would be under actual field conditions. This has led to the concern that the laboratory data taken under accelerated conditions might not be applicable to field conditions. Assuming that the environmental conditions used in the laboratory are within the range of extrapolation, a more appropriate concern would be that the laboratory data are obtained under idealized conditions such as constant stress, constant temperature, or constant chemistry. Laboratory data for Alloy 600 have been obtained using specimen geometries, which are not representative of tubing. Metallurgical conditions and chemistry conditions either have been equivalent, or have bracketed field conditions that are not known with high accuracy. Recently, data have been obtained for internally pressurized tubing in representative environments for both axial and circumferential cracking of mill annealed Alloy 600 [13]. These constant conditions and idealized geometries make it easier for the investigator to study the phenomenon, but they make it difficult to transfer data to field experience. Although difficult, the transfer is not impossible.

The most important source of differences between laboratory and field measured crack growth rates most likely does not lie with the metallurgy or the chemistry of the system. It is more likely that the differences are related to the nature of the initiation and crack growth processes taking place in an evolving environment. That is, the variations in the temperature, operational stresses, and chemistry under field conditions have an effect. Stress may be especially influential. For field conditions, the stress is not known accurately, and the stress can evolve with both time and the extent of cracking in ways that are difficult to predict. An unpressurized tube will be placed in service containing residual stress from fabrication and installation. At a minimum, there will be a short-range (~0.01 mm) tensile stress on the surfaces, especially the outside surface of the tube. There can also be stresses from roll transitions, expansions, and bends, as well as the "denting" process mentioned above. During operation, a tensile stress from pressurization will be superimposed on these residual stresses. That is, there will be a uniform base tensile stress field from the pressurization, a residual surface tensile stress that can equal the yield stress but decays over a few tenths of a millimeter, and possibly a bending stress that would add to the tensile stress on one side of the tube and subtract from the other side of the tube. In this stress field, multiple cracks can initiate and propagate into the tube wall and into a stress field that usually is decreasing.

Another concern about laboratory data and field results has to do with the number of cracks present at a location. Laboratory studies are done typically on a single crack. In actual steam generator tubes, there can be multiple initiated cracks at a position on the tube. These cracks will grow independently until they link. Once they link, growth can accelerate at the same stress level because of the dependence of the crack growth rate on stress intensity and consequently crack length. The net effect is that crack growth in the field can exhibit a stronger apparent dependence on the stress intensity factor than single cracks initiated and grown in the laboratory. Linking of cracks in the field can cause the stress needed for continued growth of the crack to approach the operating stresses on the tube. It is thought [12] that linking of cracks occurs when the depths of the cracks approach 0.5 mm, which is remarkably close to the 40% through-wall distance considered in the older repair criterion.

What emerges from these considerations is the existence of two classes of stress corrosion cracks. In the first class, there would be cracks that initiate as has been discussed and begin to grow and link. But, in this case, the loss of tensile stress as the crack propagates into the tube wall is sufficiently steep that growth cannot be sustained by the increase in the stress intensity as the cracks link. These cracks will remain arrested until something happens to increase the crack length. The regions of intergranular attack noted in Figure 1 are regions of this type. In the second class of cracks, the increase in stress intensity associated with crack linkage offsets the loss of stress as the crack passes into the tube wall. Such cracks will grow under the effect of the pressurization stress alone. This may well explain why cracks are undetected at the end of one operating cycle, but enlarge and grow through the tube wall in the next operating cycle.

4. Contentions

Contention: **Stress corrosion crack initiation and growth are not well-enough understood to be predictable in the steam generator environment. It is impossible to predict the state of degradation of the steam generator at the end of a cycle even if the state of degradation at the beginning of the cycle were well known.**

Based on the discussions in the previous subsection, it is evident that the prediction of crack growth or the increase in the voltage signal produced by a crack in the steam generator environment is a complicated process. The process for a single crack is certainly not linear. The process of assuming that the difference in voltage signals between two measurement times can be used to predict the voltage signal after another period of time will result in a significant degree of uncertainty. The assumption of a linear growth rate during a cycle is not consistent with the current mechanistic understanding of the growth process. But, an overall bounding of the process by a linear model is possible if the magnitude of the needed conservatism is tolerable. Even with this bounding, there will be extreme cases where a number of smaller cracks, barely or not detected in an earlier inspection, will link and grow at a much higher rate than would be predicted by even a bounding linear model.

Contention: **Laboratory crack growth rate measurements are not representative of the field conditions and cannot be used to predict field performance.**

Again, based on the discussions above, it is clearly challenging to relate laboratory studies of crack growth rate to field conditions. Laboratory data do span the range of stress conditions, material microstructure, and ambient chemistry that may exist in the field. Although the chemistry in the field is hard to determine, it can be argued that the chemistry in the region of importance (i.e., the crevice regions, especially for tube support plates) is within about the same throughout the fleet of affected steam generators. The concentration factor in the crevice regions is very high, and will result in aggressive chemistry in all cases, especially for drilled tube support plates. The stress states of tubes are difficult to know with any accuracy, and change during the life of a plant. The consequence of this is that there will be an uncertainty in the conditions and the data that ought to be applied to the analysis of steam generator tube performance. This will lead to a significant uncertainty in estimates of the crack growth rates and, consequently, predictions of the changes in distribution of voltages over an operating cycle will be uncertain.

> **Contention:** **Cracks that form within the region of a tube bounded by the tube support plate can extend outside of this region and not be confined fully by the tube support plate.**

The contention that cracks within the region bounded by the tube support plate can extend outside the region bounded by the plate is true, based on empirical evidence that cracks can and do extend up into the sludge pile that can accumulate on the tube support plate. There is no evidence that the cracks can grow beyond the limits of this sludge pile. Outside the confines of the crevice created by the tube support plate, the ambient chemistry of the tube is quite different than within the crevice. The environment outside the crevice is much less corrosive. In addition, tensile stress created by the accumulation of corrosion products in the crevice region declines rapidly with distance from the crevice. Alloy 600 is a very ductile material, and is resistant to unstable crack propagation in the absence of an aggressive environment where stress corrosion cracking is active. Thus, when a stress corrosion crack grows out of its aggressive environment and crack growth must occur by mechanical tearing, further crack growth will be extremely difficult. Nevertheless, it is not possible to argue that it is physically impossible for a crack to grow from the crevice region to a substantial distance beyond the tube support plate. Cracks that do grow outside the region bounded by the tube support plate are excluded from coverage by the alternative repair criteria, and must be reported to the NRC. The essential question, then, becomes an issue of the confidence with which such cracks can be detected. This issue is discussed further in Section D of this Chapter.

> **Contention:** **Laboratory burst tests are not representative of field conditions. Laboratory burst data are taken at room temperature and there is little justification for the correction of the data to actual tube conditions during operations. Crack morphology from laboratory tests are significantly different than crack morphology from actual tubes. Laboratory data ought not be used in condition monitoring correlations.**

The Ad Hoc Subcommittee concluded that there is a substantial database relating the voltage signal produced by a crack and other attributes to burst pressure. The relationships among these crack-related variables and burst pressure have been well established. Moreover, there is adequate experience to allow extrapolation of room temperature data to elevated temperatures. The extrapolation has been validated using data from Belgium [14].

The DPO author contends that the surface roughness of cracks in the laboratory data is significantly different than surface roughness of cracks in actual tubes. The DPO author is correct with respect to machined "cracks" in tubes used for some part of the database in the past. The database now contains an extensive set of data using actual stress corrosion cracks produced in the laboratory. These cracks have surface roughness that appears to be very similar to that in tubes extracted from steam generators. Interestingly, the scatter in the burst data for tubes with actual cracks is quite similar to the range of results obtained with machined "cracks." The Subcommittee did not find support for the contention concerning crack roughness.

The DPO author correctly notes that laboratory database does not contain cracks that have ligaments that will bridge sections of the crack and would thus hold the crack together and restrict leakage. Cracks in actual tubes from steam generators can have these ligaments, and the database includes such cracks with ligaments obtained by testing tubes extracted from steam generators. The industry seems to be aware of concerns about such ligaments in cracks tested for the database. They exclude from the database tests of cracks that have two or fewer uncorroded ligaments in shallow cracks less than 60% through the wall thickness. This eliminates from the database cracks that would yield unrealistically low leakage rates, but would exhibit high leakage under the conditions of a main steamline break. Although the laboratory database may lack cracks with ligaments, the total database is not so deficient. As noted in Section D of this Chapter, data from laboratory-generated specimens seem to mesh well with data obtained from tests of tubes removed from steam generators. The Subcommittee was unable to establish any significance to the contention concerning ligaments in the cracks.

> **Contention:** **The relationship between voltage and crack depth is purely empirical and tenuous at best. The relationship is not unique. That is, cracks of a variety of depths can produce the same voltage. The rate of change of voltage from an indication is not uniquely related to the rate of change of crack depth.**

The Ad Hoc Subcommittee found this contention to be true. The sensor used to measure voltage is sensitive to the volume of defects at a location, and not to the details of geometry such as crack depth and length. The ability to develop an empirical correlation between crack depth and voltage is indicative of some similarity between crack depth and crack volume that will be, at best, variable. Thus, one must be careful to restrict use of such a correlation to cracks with similar characteristics. To the extent that a correlation is used to define the criteria for repairing or removing tubes from service with cracks in the tube support plate region, additional constraints are needed to ensure that the cracks have the characteristics assumed in the correlation. It appears that GL 95-05 has such additional constraints. Cracks are to be predominantly axial, and are not to extend beyond the limits of the tube support plate. Even with these precautions, the weak correlation between voltage and crack depth means that, occasionally, short cracks of significant depth will not be detected.

It is also important that the correlation be used for interpolation with consideration of uncertainties. Such uncertainties seem to have been borne adequately in mind when defining the limiting voltage for repair. Voltages as high as 20 volts have been proposed for the limit, but in view of the uncertainties, the staff has limited the acceptance value to 1 and 2 volts for 3/4" and 7/8" tubes, respectively. Detracting badly from the empirical correlation is the substantial difference in the quality of the correlations for 3/4" tubes and for 7/8" tubes. Correlations of leakage as a function of voltage for 7/8" tubes are very much inferior to those for 3/4" tubes. Leak rate data span about three orders of magnitude for voltage within the range of interest, making correlations of very limited reliability. Similarly, the probability of leakage for 7/8" tubes seems to poorly represent the available data. The Ad Hoc Subcommittee was unable to find a phenomenological reason for the differences in these two correlations. On the other hand, the correlations of burst pressure with voltage for 3/4" and 7/8" tubes seem to be adequate for the purposes of the condition monitoring program.

The correlations with voltage are empirical. There is scatter in the correlations. Consequently, very significant caution is called for in using voltage measurements at different times to infer how the voltage distributions will change over an operating cycle. This is especially so, since the linearity assumed in this process is at odds with the mechanistic understanding of the ways that cracks grow. The Subcommittee found that the uncertainties involved in this process have been identified and addressed. The Subcommittee notes that it is still possible for behavior outside expectations to occur even when 95% confidence bounds are applied. It will be imperative to continue to develop the database used for the correlations and to reduce the uncertainties in the process. The Subcommittee feels that the DPO author has pointed out a key area of concern that must be kept in mind.

> **Contention: Because cracks can grow outside the bounds of the tube support plate, they will not be reliably constrained from rupturing by tube support plate.**

Evidence of crack growth outside the limits of the tube support plate shown to the Ad Hoc Subcommittee indicated that the growth was limited to the depth of the sludge layer that had accumulated on the top of the tube support plate. Interpretations done by the Subcommittee suggested that the chemical environment above the support plate and its sludge layer were not conducive to cracking. Stresses that could drive continued cracking appear to fall below necessary thresholds above the tube support plate. Data provided by investigators from Argonne National Laboratory suggested that axial cracks less than 1" long were unlikely to cause tube bursting for MSLB accident conditions. The Subcommittee was unable, then, to identify support for the contention that cracks would grow well beyond the tube support plate, or the limited growth above the support plate would lead to tube bursting under accident conditions without movement of the tube support plate. If the tube support plate is lifted by the forces accompanying depressurization in a main steamline break, cracks no longer occluded and restrained by the tube support plate could grow and the tubes could burst. In this regard, it is important to remember that the data on tube burst pressure is obtained without the tube support plate present. As long as moving the tube support plate does not cause additional damage to the tubes, the probability of tube bursting is still applicable even if the tube support plate is displaced.

D. Nondestructive Examination Methods and Analyses

1. Instrumentation and Repair Criteria

When wastage of the steam generator tubes was the predominant concern, the only examination required was a volumetric examination of the tubes for tube wall thickness. Eddy-current probes capable of performing reasonably reliable volumetric measurements have been available for more than 40 years. When other corrosion processes were discovered, specialists in nondestructive examination developed new, more responsive detecting elements and techniques. The first development was to use "mixed-frequency" bobbin coils to cancel out the interference caused by noise and adjacent supporting structures such as the tube sheets, tube support plates, and anti-vibration bars. Although these mixed-frequency methods provided better detection, they are poorly suited for determining the size of flaws in the tubes and are susceptible to false indications.

The next generation of detectors included rotating pancake probes and mixed array probes. These devices provided greater sensitivity for circumferential cracks and greater resolution than the mixed-frequency bobbin coils. They may not be as sensitive as bobbin coils for axial cracks. Examinations with these new devices take substantially longer than examinations with bobbin coils. As a result, these more sensitive devices are used to confirm and characterize flaws found by the mixed-frequency methods. Steam generators are still surveyed for flaws using the mixed-frequency bobbin coils.

In aging steam generators, the vast majority of flaw indications are located in the tube segments where the tubes pass through and are bounded by the tube support plates. Aggressive corrosion could take place in the crevices between the tubes and tube support plates where the impurities were concentrated. In a typical pressurized water reactor steam generator, there can be more than 40,000 sites where this aggressive corrosion environment could develop. Fortunately, stress corrosion cracking of Alloy 600 is quite sensitive to the temperature of the aggressive chemical environment, (see Section C.) Therefore, the highest rates of stress corrosion cracking occur at the hot leg entrance to the steam generator. Data provided to the Ad Hoc Subcommittee indicate what appears to be an exponential decay in the number of indications of flaws at tube support plates along the hot leg of a steam generator. Several hundred indications of flaws in tubes may be found at the first two tube support plates, less than a hundred indications in regions of the next two tube support plates, a few tens of flaws in the next support plate region, and very few flaws in the region of the rest of the tube support plates on the hot leg side. Only very occasional indications of flaws are found on the cold leg side of the steam generator.

When tube wastage was the concern, it was required that tubes have the capability to withstand pressures three times the normal operating pressure difference between the primary and secondary sides of the reactor coolant system. The tube also had to withstand the primary reactor coolant system pressure since, during a main steamline break, the differential pressure across the tube could be this large. The standard was taken to be 1.4 times the primary coolant system pressure to account for variability in tube wall thicknesses, material properties, and other uncertainties. Analyses indicated that tubes having only 40% of the original wall thickness could meet these criteria. But, an allowance for a 10% error in the measurement of the wall thickness and a 10% allowance for the continued wastage of the tube wall during the next operational cycle meant that tubes were repaired or removed when only 60% of the original wall thickness remained. With this limit, it was thought that the tubes in the steam generator might leak during a main steamline break, but they would not rupture.

A tube with a crack of some depth in its wall will be stronger than a tube with wastage to the same depth. Furthermore, tubes with cracks in the tube support plate region will be constrained against rupture by the tube support plate. Failure during accidents rather than during normal operations becomes a primary concern for tubes cracking in the regions of the tube support plates. Under the conditions of a main steamline break, blowdown forces could cause the tube support plate to move above or below the cracked region. The cracked regions would then be subject to the primary reactor coolant system pressure without external support, since the pressure on the secondary side is taken to be atmospheric. The possibility of tube rupture during a MSLB accident has to be taken into account. The standard of 1.4 times the maximum primary coolant system pressure is still applied to tubes with flaws in the region of the tube support plate.

As flaws on the outside diameter of the steam generator tubes became more numerous, a method was needed to rapidly screen flaw indications at tube support locations that would not lead to significant increases in the probability of SGTR events under normal and accident conditions. The proposed method was to use the high-speed bobbin coil probes to examine the tubes, and tests and experiments to establish a threshold bobbin coil output voltage below which further examination of individual flaws would not be required. A higher bobbin coil output voltage limit was also established to indicate flaws that had to be repaired or the tube removed from service. These alternative repair criteria were to be applied to those tube locations bounded by and not exceeding, by a specified amount, the tube support plate.

Both the NRC staff and the nuclear industry have recognized that when bobbin coil voltage limits are used as repair criteria, it was possible that some tubes would be left in service that might leak under certain accident conditions. In order to ensure that radioactive material release is within the limits specified in 10CFR Part 100, this tube leakage needs to be taken into account to establish a more accurate assessment of the risk consequences of the alternative repair criteria.

2. Prediction of Leak Rate in a Main Streamline Break Accident

The prediction of the leak rate from tubes in a steam generator during a MSLB accident involves steps shown schematically in Figure 4. The process begins with a determination of a distribution of the number of flaws found during inspection of the steam generator as a function of the magnitude of the voltage signal produced by these flaws. The number of flaws in each voltage interval is increased by a factor equal to the reciprocal of the probability of detection (1/POD) of a flaw capable of producing a signal of this magnitude. The resulting distribution is multiplied by the probability that flaws producing signals of a given magnitude will leak. The resulting distribution is then multiplied by the leak rate in each voltage range found from a correlation of the leak rate with signal magnitude. Summation over the various voltage intervals yields a prediction of the leak rate during a MSLB accident.

Contentions that have arisen concerning the various elements of this process are discussed in the subsections that follow.

LEAK RATE METHODOLOGY

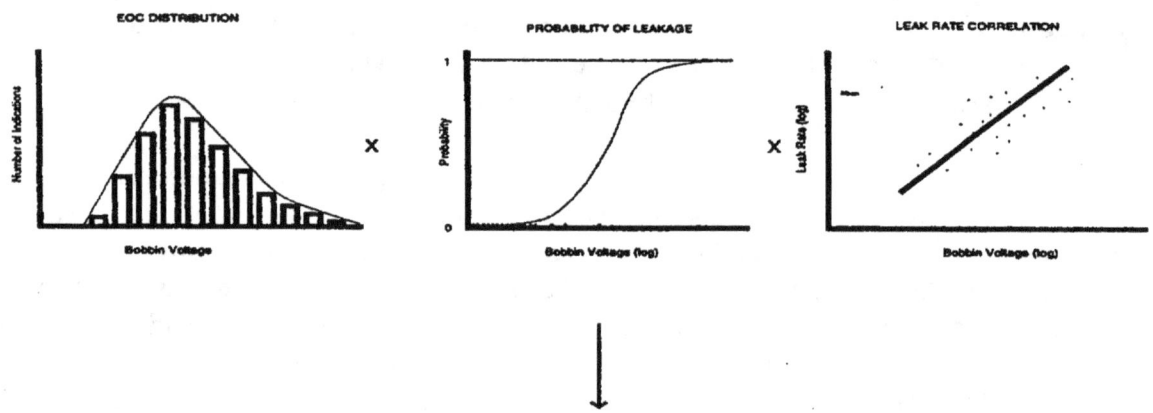

Postulated MSLB Leakage

Figure 4. Process Used to Estimate Leakage Rate

3. Probability of Detection

Contention: The probability of detection recommended by the staff is not defensible.

The staff has adopted a probability of detection (POD = 0.6) that is independent of the magnitude of the voltage signal produced by a flaw. All of the available data produced by licensees suggest that the probability of detection of a flaw increases as the voltage signal produced by the flaw increases. Some data indicate that the probability of detection approaches 0.9 for signals greater than 1 to 2 volts. Data available to the Ad Hoc Subcommittee suggest that the constant probability of detection adopted by the NRC staff is nonconservative for flaws producing voltage signals less than about 0.7 volts. This nonconservatism is of limited concern as long as low-voltage signals can be confidently ascribed to shallow, tight cracks. The constant probability of detection is quite conservative for flaws producing voltage signals in excess of about 1.5 to 2.0 volts.

On the other hand, the probability of detection adopted by the staff is a multiplier applied to the number of flaws that have been detected. If there are no flaws detected in a particular range of voltages, there is no adjustment in the distribution to reflect the possibility that undetected flaws actually do exist in the steam generator that are capable of producing signals in the particular voltage range. This deficiency is an issue only at very high and very low voltages. Otherwise, voltage intervals appear to be chosen so that they are well populated between the extremes.

The staff appears to be relying on a compensation of errors in the process to produce a result that is, overall, conservative. Based on the examples shown to the Subcommittee, the staff-approved procedure is yielding conservative results. The Subcommittee did find that the staff-approved procedure does not readily admit likely improvements in the technologies used to inspect steam generator tubes. A procedure more amenable to change with improving technology would consider the probability of flaw detection to be a function of the magnitude of the signal the flaw could produce and would adjust the number distribution to account for the probability of flaws being present that are capable of producing voltage signals in ranges where none had been found.

4. Log-logistic Probability of Leakage

Contention: The log-logistic curve is not a conservative description of the probability of leakage

There is not a unique relationship between the magnitude of the voltage signal produced by a flaw in a tube and the leakage possible from this flaw. Instead, a probability of leakage is assigned to a flaw according to the magnitude of the signal it produces. The database for developing this probability distribution consists of tests in which leakage was either detected or not detected. For 3/4" tubes, most flaws producing signals of less than 1 volt do not leak. Nearly all flaws producing signals greater than 6 volts do leak. Flaws producing signals between 1 and 6 volts may or may not leak. This database is proprietary, so it is not shown here. Although the database is growing, it will not soon be large enough to develop confidently an empirical probability distribution. Instead, Westinghouse has proposed and the staff has accepted a hypothesized distribution. A log-logistic distribution has been proposed:

$$\text{Probability of Leakage} \ = \ \frac{1}{1 + e^{-(a + b\log V)}}$$

where V is the voltage signal produced by the crack or flaw. The parameters "a" and "b" in the distribution are adjusted to approximate the data well. There is no implication that this distribution has any fundamental, phenomenological relationship to leakage from flaws. It is only argued that the distribution can adequately reflect the available database.

Buslik [15] has examined this proposed distribution in detail and has found that many features of the hypothesized distribution would be found in other plausible distributions for the probability of leakage. The effect of the distribution is first to attribute some small probability of leakage to flaws producing very low-voltage signals. For example, a version of the distribution that has been used recently indicates there to be about a 1% probability that a flaw producing a 1-volt signal will leak. Alternative distributions would yield different, but still small, probabilities for leakage from flaws producing such low-voltage signals. In light of the scatter in the data, such discrepancies in the predictions of the very low probabilities of leakage from low-voltage signals hardly seem a basis for judging the adequacy of the log-logistic distribution relative to other continuous distributions that could be hypothesized. On the other hand, the distribution also leads to the prediction that there is a small probability that flaws producing very large signals will not leak. For instance, the same distribution that yields a 1% probability that a 1-volt flaw will leak predicts a 25% probability that a flaw producing a 5-volt signal will not leak. Indeed, the database indicates that there are instances in which 3/4" tubes with flaws producing 5-volt signals did not leak during testing. The probability distribution predicts that there is about a 3% probability that a flaw producing a 10-volt signal will not leak. There are no instances in the database in which a flaw producing such a large signal did not leak, but the database is still small. As the database is expanded, there might be a case in which a tube with a flaw producing a signal larger than 10 volts does not leak.

Once the procedure involving the probability of leakage distribution is accepted, the only criterion for judging the adequacy of the hypothesized log-logistic distribution is whether it plausibly reflects the database. Special attention is, of course, needed for the extremes of the distribution. Comparison of the data to the proprietary database for 3/4" tubes does suggest a plausible reflection of the database. On the other hand, comparison of the distribution to the database for 7/8" diameter tubes is more questionable. Data are more scattered for these tubes that are only modestly larger. The distribution indicates about a 25% probability that flaws producing 20-volt signals will not leak and about a 3% probability that flaws producing a signal of 100 volts would not leak.

The Ad Hoc Subcommittee concluded that in the case of the 7/8" tubes, there was a need to end reliance on a completely continuous distribution for the probability of leakage, and that some engineering judgment should be introduced, especially for flaws producing signals in excess of about 30 volts. (Of course, tubes found to have flaws producing voltage signals this large would be repaired or removed from service.)

5. Correlation of Leakage with Voltage

Contention: The relationship between voltage and leak rate is not accurate enough to be used. A correlation between voltage and leak rate does not exist.

The Ad Hoc Subcommittee examined proprietary data used to formulate a correlation between leakage under main steamline break conditions and the voltage signals generated by flaws in tubes. A distinction between the database for 3/4" tubes and the database for 7/8" tubes needs to be made to ensure that in the case of 3/4" tubes, there was more scatter in the data than would have been desired. There was evidence, however, of a linear correlation between the logarithm of the voltage and the logarithm of the leak rate. Scatter in the measured leak rates about the correlation line amounted to more than an order of magnitude. Much of this scatter was in the direction of lower leak rates for a given voltage. Statistical analyses used to define conservative predictions of the correlation appeared appropriate.

Leakage data for 7/8" tubes did not appear to correlate at all well with the voltage signal produced by flaws in the tested tubes. Although statistical analyses indicate some sort of correlation, the Ad Hoc Subcommittee concluded that this correlation was tenuous, at best. In the range of voltages that is relevant to the application of the alternative repair criteria, measured leakage rates span a range of three orders of magnitude. Any use of the supposed correlation to predict leakage should account for a very high degree of uncertainty in the data. Clearly, there is a very real need to further develop the database for 7/8" tubes. The Ad Hoc Subcommittee was unable to identify any phenomenological reasons for there being so much worse a correlation of voltage and leak rate for the 7/8" tubes than for the 3/4" tubes.

Even though the data for tube leakage and flaw voltage may be poor, these data do emphasize that tubes can leak. The aggregate tube leakage can be estimated for plant accident conditions and measured for normal operating conditions. The Subcommittee notes that this is a better situation than what previously existed when there was no knowledge about the nature and extent of steam generator tube leakage during accident conditions. Then, only assumed regulatory limits could be used to estimate accident events and radionuclide releases.

6. Prediction of the Probability of Tube Burst

A schematic diagram of the process used for estimating the probability of tube burst is shown in Figure 5. The end of cycle distribution of flaws corrected for the probability of detection and the growth of flaws over the cycle is used just as in the estimation of leakage. This distribution is convoluted with the correlation of normalized burst pressure with voltage. The normalization factor is the flow stress, which is the average of the yield stress and the ultimate stress. The flow stress has a distribution due to the variability of material properties, so the analysis is done using Monte Carlo methods. The Ad Hoc Subcommittee did not examine the distribution of the flow stress. It did examine the correlations of burst pressure with voltage.

Proprietary databases for the burst pressure as a function of the voltage signal produced by a flaw in a tube segment are available for both 3/4" and 7/8" tubes. The databases have been formed using data from tubes removed from steam generators, tubes cracked in laboratory studies, and tubes with machined "cracks." Data are available for voltage signals of about 0.1 to 100 volts. Lower bound burst pressures vary by about a factor of four over this voltage range. Scatter in the data is significant. For both sizes of tubes the scatter in the burst pressure amounts to about 20% of the range of the data. Linear correlations of the burst pressure with the logarithm of the voltage have been developed.

Some contentions concerning the database have been discussed in Section C of this Chapter in connection with the discussion of stress corrosion cracking. Other contentions are discussed here.

Contention: Removing tubes from service damages the flaws so they cannot be used to develop a reliable correlation of the burst pressure with voltage.

It is true that removing a tube from service is a vigorous undertaking that can cause additional damage to flaws being tested for burst pressure. In general, voltage indications after removal of a tube from service are no less and often greater than the voltage indications prior to tube removal. The Ad Hoc Subcommittee was unable to identify plausible mechanisms for a flaw to be partially healed by the removal process, so it took the increase in voltage to indicate that a flaw had been further damaged by the removal process. The correlation of burst pressure against the logarithm of the voltage is based on the voltage prior to removal. This should introduce additional conservatisms in the database that are not reflected in the usual statistical analysis of scattered data. As correctly noted by the DPO author and discussed earlier in subsection A.1.a, Damage Progression, of this Chapter, the leakage database may not be conservative if blowdown forces extensively damage and rupture ligaments in ways not reflected by the testing. Pending resolution of the damage progression issue, the Subcommittee felt burst data obtained from tubes removed from service should be used in developing the burst pressure correlation with voltage.

The Ad Hoc Subcommittee also examined the database in terms of the sources of the data. Data obtained from tubes flawed in model boiler tests constitute much of the data available for higher voltage signals. These data mesh smoothly with the lower voltage data obtained from tubes removed from service. That is, there was no readily apparent indication that the two data sources yielded results that were indicative of radically different population distributions for flaws.

CALCULATING THE PROBABILITY OF RUPTURE

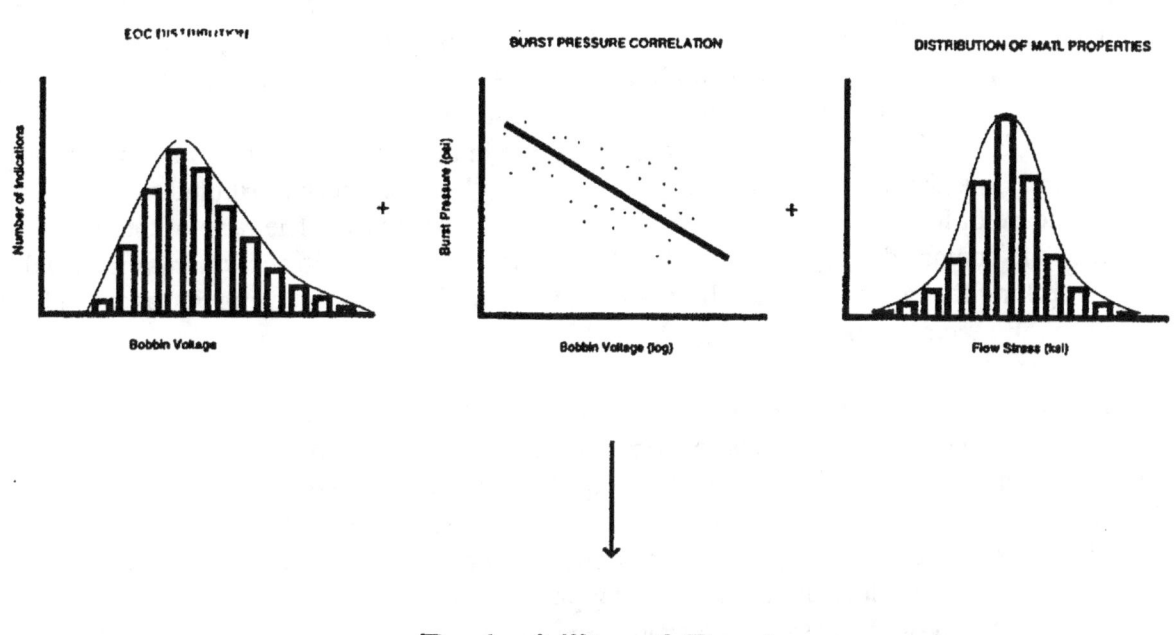

Probability of Rupture
Given a MSLB

Figure 5. Process Used to Estimate the Tube Burst Probability

E. Iodine Spiking and Source Term Issues

Contention: **The iodine spiking factor used for accident consequence analysis at plants with iodine coolant concentrations limited to less than 1.0 µCi/g and adopting the alternative repair criteria is too low.**

During normal power plant operation, there is a steady-state concentration of radioactive iodine in the coolant. In the past, when cladding on reactor fuels was less reliable, the steady-state concentration of iodine was thought to be due predominantly to the escape of iodine from defected fuel rods. Technical Specifications for many plants restrict the allowable concentration of radioactive iodine in the coolant to less than 1.0 µCi/g

Analyses of the consequences of MSLB and SGTR design basis accidents focus on the release of coolant contaminated with radioactive iodine from the primary system through the secondary system and into the environment. It has been observed that sudden reductions in reactor power and large pressure differentials between the primary and secondary systems, which would be expected to occur during MSLB and SGTR accidents, lead to increases in the concentration of radioactive iodine in the coolant. The phenomenon is called "iodine spiking" and has been attributed to increases in the rate of release of iodine to the coolant from defected fuel rods [16]. The ratio of the release rate of iodine to the coolant during an accident to the release rate during normal operations is called the "iodine spiking factor." Most plants have analyzed the MSLB accident using 500 as the value of the iodine spiking factor recommended by the staff.

Today, typical concentrations of radioactive iodine in the primary coolant system are many times lower than the Technical Specification limit. It is difficult to attribute the very low steady-state concentrations of iodine in the coolant to leakage from defected fuel rods. It has been suggested that the steady-state concentrations observed today come predominantly from fissioning of tramp fuel particles in the reactor coolant system. Such a source for the steady-state iodine concentration would not be expected to produce the iodine spiking phenomenon in the event of a sudden reduction of reactor power or a drop in system pressure. Presumably, this suggestion could be confirmed by monitoring the concentrations of other fission products in the coolant.

When the alternative repair criteria are adopted, the potential leakage of contaminated coolant from the primary system can be orders of magnitude greater than the limit of 1 gpm assumed in the past. In order to meet the 10 CFR Part 100 dose criteria and limits on the dose to operators during a hypothesized accident, licensees have proposed to reduce the Technical Specification limit on the radioactive iodine concentration in the coolant during normal operations from 1.0 µCi/g down to 0.01 µCi/g. This limit is still well above iodine concentrations during normal operations at most nuclear power plants. Licensees propose to continue to use the iodine spiking factor of 500 for the analysis of accident consequences.

The DPO author contends that the spiking factor used for the accident analyses is too low when the Technical Specification limit on the iodine concentration in the coolant during normal operations has been reduced. He argues that the spiking factor increases with decreasing steady-state iodine concentration. Data collected and reviewed by Atwood and Sattison [17] substantiate the claim made by the DPO author (See Figure 6). A phenomenological understanding of this correlation between the iodine spiking factor and the steady-state iodine concentration in the coolant is not well established.

A more extensive database on iodine spiking has been published by Adams and Atwood [18]. These authors did not cast the database in terms of an iodine spiking factor nor did they provide the information necessary to cast the data they published in this form. The DPO author and the staff have been able, apparently, to convert the more extended database into iodine spiking factors. The more extended database indicates very high iodine spiking factors for low initial concentrations of iodine. Unfortunately, the database includes instances in which the iodine spiking factor is less than unity which is inconsistent with the hypothesized mechanism for the iodine spiking. The Ad Hoc Subcommittee feels that the extended database cast by some means into the terms of iodine spiking factor is indicative of:

- increased uncertainty in measurements at low concentrations, or

- a possible change in the mechanism for the steady-state concentration of iodine in the coolant during normal operations that does not lead to large iodine spikes under accident conditions.

In any event, the Ad Hoc Subcommittee has chosen to examine the database published by Adams and Sattison because of the more certain origins of iodine spiking factors published in this paper. The Subcommittee does not presume that its analyses are definitive, since they do not include the entire database that is now available. The analyses are presented here simply to provide some indication of the magnitudes of iodine spiking factors and their dependencies in addressing the contentions concerning the magnitude of the spiking factor to use in design basis accident analyses.

Few steam generator tube ruptures have occurred. The databases cited above are predominantly for reactor trip events, which are thought to produce iodine spiking that is similar to that produced during steam generator tube ruptures. Adams and Atwood [18] have argued that peak iodine concentrations reported by licensees for reactor trip events should be multiplied by a factor of three to account for delay in making the measurements and radioactive decay. This, they contend, will yield bounding values of the iodine spiking factor. The Ad Hoc Subcommittee has accepted this argument, but has found it necessary to multiply the spiking factor rather than the peak iodine concentration to bound the data properly since insufficient data were published to follow the recommendation by Adams and Atwood.

Unweighted least-squares regression of the iodine spiking factor data published by Adams and Sattison and multiplied by three against the initial steady-state iodine concentration yields:

$$\log_{10}(SF) = 1.266(\pm 0.347) - 0.891(\pm 0.299)\log_{10}[C]_{ss}$$

where:

SF = iodine spiking factor

$[C]_{ss}$ = steady-state iodine concentration in the coolant (μCi/g).

The linear correlation coefficient for this expression is only about-0.77. The correlation provides a prediction of the mean of the uncertain spiking factor at a given steady-state iodine concentration. Because the data set is finite, the mean is uncertain. The 95th percentile confidence interval[5] for the predicted mean from this correlation is found from:

$$\log_{10}SF_{97.5/2.5} = \log_{10}SF \pm 2.00 \times 0.322 \left[\frac{1}{N} + \frac{(\log_{10}[C]_{ss} + 1.075)^2}{11.219}\right]^{1/2}$$

where $N = 58$ is the number of data points used to construct the correlation. These bounds are indicated in Figure 6 by the dashed lines on either side of the solid line calculated from the correlation.

The 95th percentile values[6] of the iodine spiking factor obtained from this correlation for various concentrations of iodine are:

$[C]_{ss}$	SF_{95}
1.0 μCi/g	28
0.1	169
0.01	1630

Adams and Sattison analyzed the data that they collected, assuming there to be a constant iodine spiking factor that depended solely on reactor trip. They seemed unaware of the correlation of the spiking factor with the initial steady-state iodine concentration. They considered the data to represent a sampling of the distributed population of iodine spiking factors. Following similar reasoning, Adams and Atwood contended that the iodine spiking factor value of 500 was a factor of about 15 conservative for the analysis of SGTR accidents. Because the apparent correlation of spiking factor with initial iodine concentration is now known, this line of reasoning can be discounted. But, and undoubtably coincidently, the above regression predicts that the spiking factor value of 500 is conservative by the same factor of about 15 for initial iodine concentrations near 1.0 μCi/g, which is the common Technical Specification limit. It would not be conservative for initial iodine concentrations less than about 0.03 μCi/g.

The staff provided the Ad Hoc Subcommittee with an expanded data set on spiking factors. The pedigree of this database is uncertain, but it definitely is not the database used to prepare plots of spiking factor as a function of the initial, steady-state iodine concentration of the coolant shown to the Subcommittee and included in the staff's responses to the DPO. This database seems to be composed of two populations (See figure 7). In one population, the spiking factor does exhibit a dependence on the initial, steady-state iodine concentration. Spiking factors from the other

---

[5] 95% confidence that the true mean lies within the calculated interval.

[6] 95% confidence that the true mean is less than or equal to the tabulated value.

population appear to be independent of this concentration. Based on this a conventional linear least squares analysis[7] of the population exhibiting a dependence on concentration:

$$\log_{10}(SF) = 1.906(\pm 0.096) - 0.355(\pm 0.054)\log_{10}[C]_{ss}$$

Based on this:

$[C]_{ss}$	SF_{95}
1.0 µCi/g	135
0.1	281
0.01	628

Completely empirical correlations of the type shown above that do not have a mechanistic underpinning ought not be extrapolated beyond the supporting database. When they are, properly calculated confidence intervals become very large. Worse, predictions can be useless because they indicate greater iodine release than the iodine inventory available for release from one or a few defected fuel rods. The Ad Hoc Subcommittee notes, then, that a simple correlation of spiking factor with initial iodine concentration is not an adequate description of the spiking phenomenon for accident analyses. There have been attempts to develop mechanistic spiking models [19] and these models could be developed further to gain a more realistic understanding of the source terms associated with design basis accidents.

The databases assembled by Adams and Sattison and by Adams and Atwood are thought to be directly applicable only to SGTR accidents. There is a widespread belief that the larger reductions in system pressure accompanying MSLB events will lead to high values of the iodine spiking factor. The Ad Hoc Subcommittee is not aware of a well-developed database to support this belief; however, the Subcommittee suspects that a database similar in quality to that discussed above might be constructable. The staff has chosen not to assemble such a database. Instead, the staff has argued that the effect of depressurization during a main steamline break can be bounded by assuming that the iodine spiking factor varies with the square of the change in pressure. Based on this hypothesis, the staff contends that the databases for reactor trip events can be scaled to the conditions of a main steamline break by multiplying the spiking factors by factors of 4 to 9. The staff has argued that this scaling is not needed because of the factor of 15 conservatism suggested by Adams and Atwood in the spiking factor value of 500. This conservatism, of course, was based on a hypothesis concerning the data now shown to be untrue.

---

[7] The Ad Hoc Subcommittee feels that the data set should be treated as though both the dependent and the independent variables are uncertain rather than by conventional least squares methods. This type of treatment yields a correlation with a stronger dependence on concentration. On the other hand, weighting the data to reflect greater uncertainty in values at low concentrations reduces the implied dependence on concentration. Clearly, a more careful examination of these data is merited.

The Ad Hoc Subcommittee was unable to identify defensible technical bases for the staff decisions to:

- not consider the correlation of the iodine spiking factor with initial iodine concentration
- accept the conservatism suggested by Adams and Atwood in discussing the analysis of accident consequences for a main steamline break, but neglect it for the analysis of steam generator tube rupture events
- not scale the database on iodine spiking factor for reactor trip events to main steamline breaks

The DPO author rightly contends that the iodine spiking factor seems to exhibit a correlation with the initial iodine concentration in the coolant. But, it might be argued that for accident analyses the appropriate spiking factor to use is that found from the empirical correlation using the actual operational iodine concentration rather than the Technical Specifications limit adopted by the licensee, whether it is 1.0 μCi/g or 0.01 μCi/g. Such a procedure would be inconsistent with the approaches that have long been used in the analysis of design basis accidents. The Ad Hoc Subcommittee also believes that the empirical, linear correlation of spiking factor with initial iodine concentration to the very low initial iodine concentrations typical of many plants today without explicitly accounting for data uncertainties and variations in the mechanisms that lead to operational levels of iodine in the coolant is without sound technical foundation.

Correlation of Bounded Iodine Spiking Factor Data

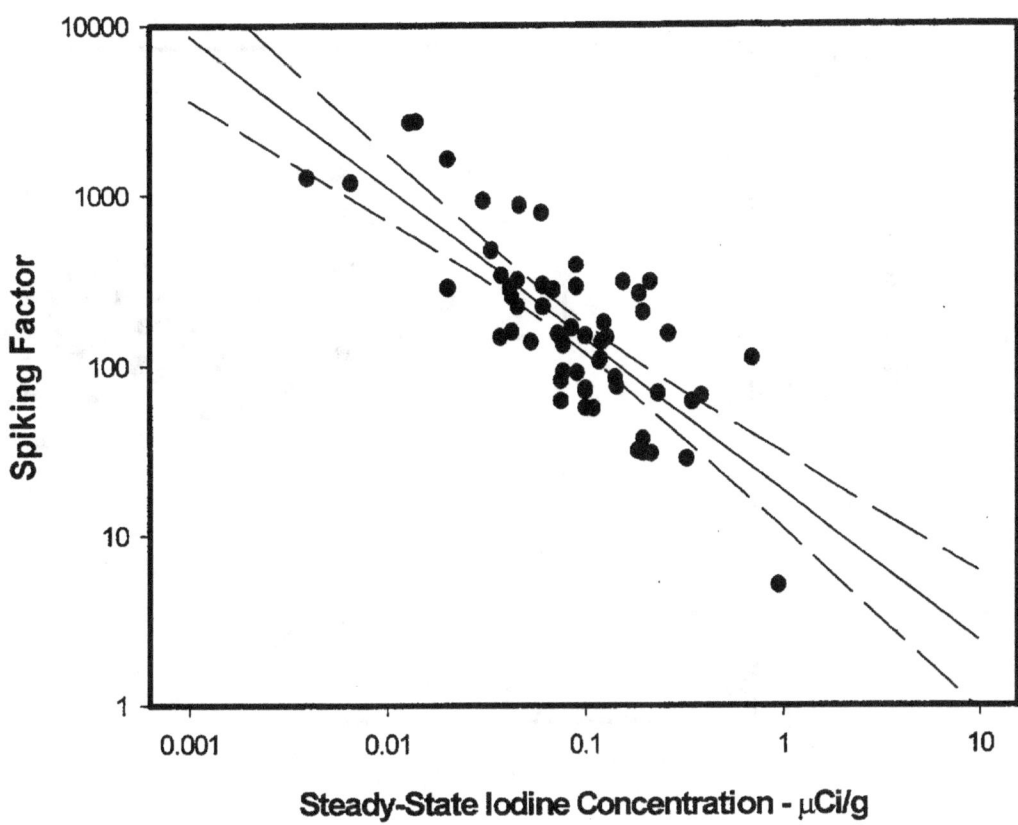

Figure 6. Correlation of Bounded Iodine Spiking Factor Data with Initial Iodine Concentration in the Coolant. Solid circles denote values of the spiking factor published by Adams and Sattison multiplied by a factor of 3. The solid line is the prediction of the correlation described in the text. Dashed lines define the 95% confidence interval for predictions from the correlation.

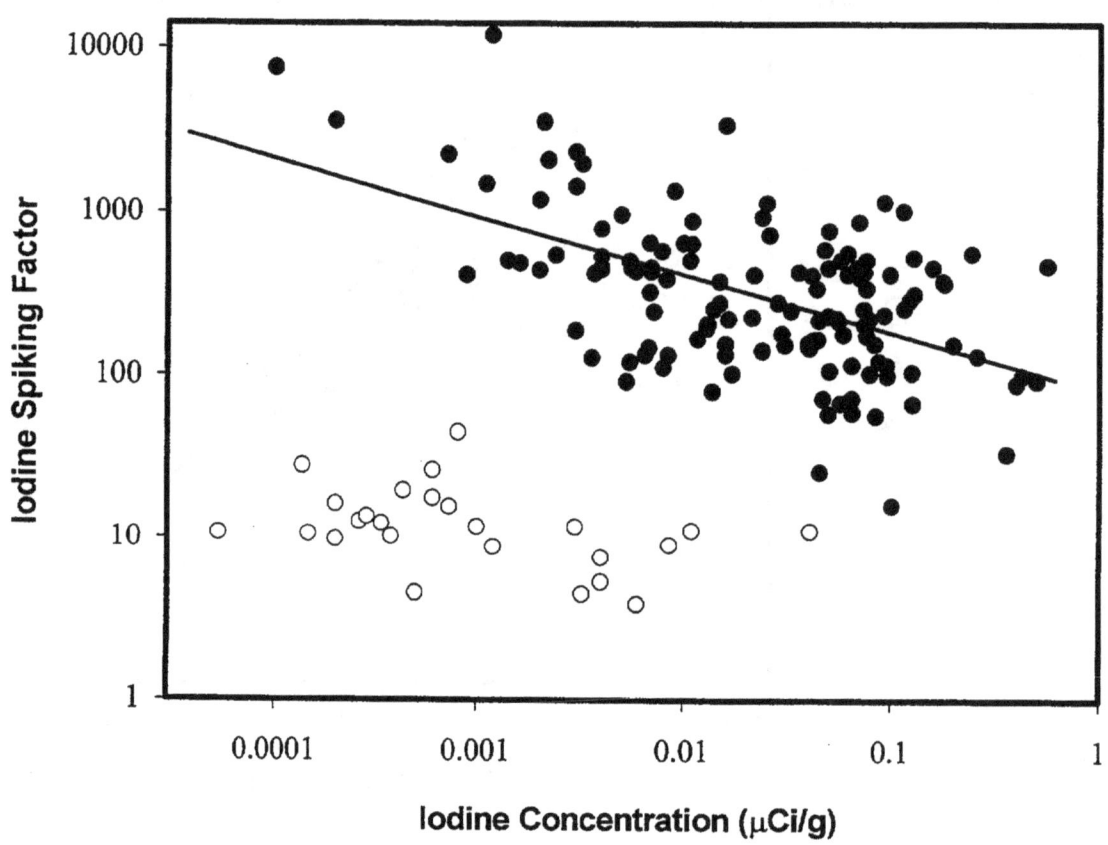

Figure 7. Expanded Data Set Showing Two Populations of Spiking Factor Data.
Filled circles come from a population that exhibits a dependence on initial iodine
concentration in the coolant. Open circles are for a population that seems independent of
this concentration. Concentration dependent data have been used to fit a line as described
in the text.

4. CONCLUSIONS AND RECOMMENDATIONS

The DPO author has raised substantive technical issues. These issues relate specifically to the alternative repair criteria in some cases and in other cases relate to the generic evaluation of the risk profile of nuclear power plants regardless of the repair criteria that have been adopted. In nearly all cases, the staff has made or is making a diligent effort to address the issues. Significant conclusions and recommendations by the Ad Hoc Subcommittee concerning these issues are summarized in this Chapter.

Conclusion: There is a need for alternative repair criteria.

There is no question that the phenomena affecting steam generators have evolved since repair criteria were first established. Consequently, different repair criteria are needed that preserve the protection afforded the public health and safety. Degradation of steam generator tubes is occurring in the annular regions bounded by the tube support plates that would be difficult to detect and characterize by any means. Technical advances have resulted in improved eddy-current techniques for detecting flaws. Such improved techniques make repair criteria based on voltage signals attractive especially if supplemented by characterizations that ensure flaws producing the signal meet explicit and implicit assumptions about the possible growth and behavior of the flaws.

Conclusion: Plants will be operated with flaws in the steam generator tubes and this need not be risk significant.

Eddy-current techniques are not capable of 100% accuracy in detecting flaws. Indeed, no method of detection can be of perfect accuracy. This does not degrade the protection afforded the public health and safety, provided the risk is properly managed. If the risk can be managed properly, it is acceptable to operate plants with known, small flaws as well as undetected flaws in the steam generator tubes. Among the measures that have been taken to manage the risk in such cases is the modification of usual analyses of design basis main steamline rupture events to include induced rupture of one or more steam generator tubes. Additional, defense-in-depth management of risk can be achieved by restricting known flaws in the steam generator tubes to those unlikely to grow significantly during an operational cycle.

Conclusion: The general features of the procedures that the staff has established to limit the number and size of flaws left in operating steam generator tubes are adequate.

The Ad Hoc Subcommittee found no fault with the general concept of a threshold voltage signal as a criterion for repair subject to constraints to ensure that the flaw producing this signal meets explicit and implied assumptions. The Subcommittee reached this conclusion recognizing that the abilities of current detection technologies are limited and it is possible that the threshold criteria would leave unrepaired flaws of objectionable size in the steam generator. The Subcommittee also found the thresholds of 1 and 2 volts established by the staff to be conservative. The Subcommittee did not attempt to reach conclusions concerning occasions when the staff granted exemptions to these criteria, except to note that these exemptions should have been accompanied by more complete risk analyses.

The Ad Hoc Subcommittee concluded that the condition monitoring program that licensees adopt in conjunction with alternative repair criteria, although not perfect, can produce a better understanding of the conditions and vulnerabilities of steam generators than in the past. These data together with proper interpretations afford additional protection to the public that has not been possible in the past. The Subcommittee did not attempt to investigate the quality with which the condition monitoring is being implemented by licensees. The Subcommittee is aware, however, of recent events that may suggest implementation does not meet expectations of the staff.

> **Recommendation:** **Risk analyses that the staff considers need to account for progression of damage to steam generator tubes in a more rigorous way.**

The DPO author suggests several mechanisms that could lead to rupture of multiple steam generator tubes. Most notable among these is damage progression as a result of the dynamic processes associated with depressurization during a main steamline break or other type of accident. The Ad Hoc Subcommittee found that the staff did not have a technically defensible understanding of these processes to assess adequately the potential for progression of damage to steam generator tubes. Bending and flexion of the tubes produce conditions regarding crack growth, tube leakage, and tube burst outside the range of analyses and experiments done by the staff. Although the Subcommittee felt that current practices associated with the alternative repair criteria and condition monitoring were sufficiently conservative as an interim measure, there is an imperative for the staff to act expeditiously to develop a much better understanding of the dynamic processes associated with depressurization and how the processes could lead to damage progression.

Similarly, the Ad Hoc Subcommittee did not feel that the staff had developed an adequate understanding of how movement of the tube support plates during an event could damage the tubes and augment leakage from the primary side to the secondary side of the reactor coolant system. The staff needs to develop an understanding of how tube support plate movement could lead to unplugging of cracks occluded by corrosion products in the annular space between the tube support plate and the tubes. Furthermore, the staff needs to ensure that the plate movement will not induce additional damage to the tubes such as causing the cracks to grow, possibly to the point of tube burst.

On the other hand, the Ad Hoc Subcommittee did feel that the staff had undertaken adequate research to address the issue of fluid jets from cracks in tubes could pierce adjacent tubes. Although it is still necessary to carry this research to an appropriate conclusion, early results suggest that damage progression by the jet cutting mechanism is not likely.

The Ad Hoc Subcommittee did not find evidence to suggest that the "unplugging" of cracks outside the annular region of tubes bounded by the tube support plate would be a damage progression mechanism of concern. The staff has developed an adequate understanding of crack growth and expansion in these regions under the conditions of depressurization.

> **Conclusion: Substantial uncertainties remain in the understanding of steam generator tube performance under severe accident conditions.**

The staff has not developed persuasive arguments to show that steam generator tubes will remain intact under conditions of risk-important accidents in which the reactor coolant system remains pressurized. The current analyses dealing with loop seals in the coolant system are not yet adequate for risk assessments. The treatments of mixing of flows in the inlet plenum to a steam generator under conditions of countercurrent natural convection flow are optimistic and are not substantiated by applicable data from experiments. Sensitivity studies have not explored the plausible ranges of parameter values or the space of uncertainties adequately. Finally, the Ad Hoc Subcommittee notes that analyses of failure of other locations in the coolant system subjected to natural convection heating have not included a systematic examination of vulnerable locations in the system.

These deficiencies constitute uncertainties in the state of knowledge concerning the progression of severe reactor accidents. There are, of course, many other uncertainties in severe accident progression. The uncertainties in steam generator performance under severe accident conditions arise regardless of the repair criteria applied to the steam generator tubes. The Ad Hoc Subcommittee was not able to quantify any incremental risk that might be associated with the alternative repair criteria relative to the 40% through-wall criterion in this regard.

Conclusion: Analyses of human performance errors during design basis accidents appear consistent with the current practices.

The low probability (10^{-3}) of human error leading to core damage in response to either a steam generator tube rupture or a main steamline break adopted by the staff appears consistent with the state of current understanding of human performance errors when only a single tube ruptures. In developing assessments of risk concerning these design basis accidents, the staff must consider the probabilities of multiple tube ruptures until adequate technical arguments have been developed to show damage progression is improbable. In all cases, the staff needs to develop defensible analyses of the uncertainties in its risk assessments, including uncertainties in its assessments of human error probabilities. As the staff develops a better understanding of the dynamic processes associated with depressurization during a main steamline break, it may want to revisit estimates of operator error probability in light of the considerable operator distraction that might occur during such events.

Conclusion: The general features of the condition monitoring program are adequate.

The Ad Hoc Subcommittee found the general features of the processes being used to assess the probabilities of leakage and probabilities of tube burst to be conservative. The Subcommittee did feel it technically defensible to develop empirical correlations of burst pressure and leakage with voltage signal. The Subcommittee found no evidence that the databases were flawed in any nonconservative, systematic ways by the use of data from tests of pulled tubes and laboratory specimens. The Subcommittee did not find a need to further develop the functional form of the equations used to estimate the probabilities of leaking and the probabilities of bursting of tubes. The Subcommittee did feel that the constant probability of detection (POD) approved by the staff suffered from deterring technical improvements, but the Subcommittee is aware that the staff will consider approving alternative descriptions of the probability of detection that recognize that the probability of detection can depend on flaw size.

Recommendation: **The databases for 7/8" tubes need to be greatly improved to be useful.**

The correlation of leakage with voltage used in the condition monitoring program for 7/8" tubes does not correspond very well with the correlation of leakage with voltage for 3/4" tubes. The Ad Hoc Subcommittee could identify no mechanistic reasons why data for the 7/8" tubes should so poorly relate to the correlations achieved with data for 3/4" tubes. The lack of relationship may reflect stochastic scatter and the limited size in the database. The staff should consider requiring a near-term expansion of this database.

Recommendation: **The staff should establish a program to monitor the predictions of flaw growth for systematic deviations from expectations.**

A step in the condition monitoring program is the prediction of the change in the voltage distribution over an operating cycle. This is done assuming a linear change in the distribution with time which is inconsistent with the understanding of the behavior of stress corrosion cracks established in the research that the staff has supported. Flaws grow slowly until they can interlink. Once they interlink, it is possible for flaws to grow quite quickly. Flaw growth is, then, inherently nonlinear. It can be treated as linear with time only in a bounding manner. Even then, stochastic variability means that occasionally individual flaws can violate even very conservative, linear bounds. The probability of stochastic violation of the bounds can be limited by the choice of the conservatism in the bounds. Of more concern would be a systematic violation of the linear bounding of the growth process. It will be important for the staff to be vigilant in monitoring the implementation of the alternative repair criteria to watch for such systematic errors in the flaw growth predictions. Therefore, a program to develop a database on predictions of voltage distributions and observed voltage distributions is needed.

Recommendation: **The staff should develop a more technically defensible position on the treatment of radionuclide release to be used in safety analyses of design basis events.**

The Ad Hoc Subcommittee has concluded that the staff has not adopted a technically defensible position on the choice of the iodine spiking factor to be used in the analyses of design basis accidents for compliance with the requirements of 10 CFR Part 100 or General Design Criterion (GDC) 19. The constant factor of 500 is inconsistent with the available data on spiking factor. Arguments that the spiking factor is adequate for analyses of main steamline breaks, because its conservatism compensates for any dependence on pressure differential, are not technically defensible. Conservatism attributed to the spiking factor is based on an incorrect interpretation of the data.

Data are available to develop a correlation of the spiking factor with steady-state iodine concentration for analysis of steam generator tube rupture events. If this is done, the correlation should be used with whatever Technical Specification limit on the iodine concentration has been adopted by a licensee rather than the operating iodine concentration typical of the plant. This will maintain consistency with past practices in the analyses. If the staff continues to believe that the spiking factor is greater in a main steamline break because of high differential pressures, another specification of the spiking factor needs to be developed. The Ad Hoc Subcommittee believes, however, that it would be better for the staff to develop a more realistic mechanistic understanding of the radionuclide release during design basis accidents and use that understanding for safety analyses. Good starts on such mechanistic analyses have been made and need to be pursued to the point that they yield products that can be reliably used in the regulatory process.

5. REFERENCES

1. U.S. Nuclear Regulatory Commission, "Voltage-Based Repair Criteria for Westinghouse Steam Generator Tubes Affected by Outside Diameter Stress Corrosion Cracking," Generic Letter 95-05, August 3, 1995.

2. U. S. Nuclear Regulatory Commission, Office of Nuclear Regulatory Research, NUREG-1150, "Severe Accident Risks: An Assessment for Five U.S. Nuclear Power Plants," Volumes 1 and 2, June 1989.

3. Nuclear Energy Institute, "Steam Generator Program Guidelines," NEI 97-06 (Rev. B), Washington, D.C., January 2000.

4. U.S. Nuclear Regulatory Commission, "Individual Plant Examination Program: Perspectives on Reactor Safety and Plant Performance, Summary Report," NUREG-1560 Volume 1 Part 1, October 1996.

5. U.S. Nuclear Regulatory Commission, "Risk Assessment of Severe Accident-Induced Steam Generator Tube Rupture," NUREG-1570, March 1998 and W.A. Stewart, A.T. Pieczynski, and V. Srinivas, "Natural Circulation Experiments for PWR High-Pressure Accidents," EPRI Report TR-102815, Electric Power Research Institute, Palo Alto, California, August 1993.

6. U.S. Nuclear Regulatory Commission, "Voltage-Based Interim Plugging Criteria for Steam Generator Tubes," NUREG-1477 (Draft Report for Comment), June 1993.

7. U. S. Nuclear Regulatory Commission, NUREG/CR-6365, "Steam Generator Tube Failures," Prepared by Idaho National Engineering Laboratory, INEL-95/0383, April 1996.

8. Idaho National Engineering Laboratory, INEL-95/0641, "Steam Generator Tube Rupture Induced from Operational Transients, Design Bases Accidents, and Severe Accidents," August 1996.

9. A.D. Swain and H.E. Guttmann, "Handbook of Human Reliability Analysis with Emphasis on Nuclear Power Plant Applications," NUREG/CR-1278, SAND80-0200, Table 20-1, Sandia National Laboratories, Albuquerque, NM, October 1980.

10. S. Majumdar, "Assessment of Current Understanding of the Mechanisms of Initiation, Arrest and Reinitiation of Stress Corrosion Cracks in PWR Steam Generator Tubing," NUREG/CR-5752, ANL-99/4, Argonne National Laboratory, Argonne, Illinois, February 2000.

11. J.C. Danko, "Corrosion in the Nuclear Power Industry" in **Metals Handbook**, Volume 13, Ninth Edition, 1987, pp. 924—984.

12. B.W. Brisson, R.G. Ballinger and A. R. McIlree, "Intergranular Stress Corrosion Cracking Initiation and Growth in Mill-Annealed Alloy 600 Tubing in High-Temperature Caustic," **Corrosion**, *54* (1998) 504.

13. D.N. Kalinousky, "Measurement of PWR Steam Generator Tube Degradation," SM Thesis, Department of Nuclear Engineering, Massachusetts Institute of Technology, September 1998.

14. R. Keating, P. Hernalsteen, and J. Begley, **"Steam Generator Management Project Burst Pressure Correlation for Steam Generator Tubes with Through-Wall Axial Cracks,"** EPRI Report TR-105505, October 1997.

15. A. Buslik, "Model and Parameter Uncertainty in the Estimation of Steam Generator Tube Leakage Probability as a Function of Eddy Current Voltage," NUDOCS Ascension Number 9408080202.

16. R.J. Lutz and W. Chubb, "Iodine Spiking — Cause and Effect," **Transactions American Nuclear Society,** *28* (1978) 649, and K.H. Neeb and E. Schuster, "Iodine Spiking in PWRs: Origin and General Behavior," **Transactions American Nuclear Society,** *28* (1978)650.

17. J.P. Adams and M.B. Sattison, "Frequency and Consequences Associated With a Steam Generator Tube Rupture Event," **Nuclear Technology,** *90* (1990) 168.

18. J.P. Adams and C.I. Atwood, "The Iodine Spike Release Rate During a Steam Generator Tube Rupture," **Nuclear Technology,** *94* (1991) 361.

19. B.J. Lewis and F.C. Iglesias, **"An Iodine Spiking Model for Pressurized-Water Reactor Analysis, Volume 1: Theory Manual,"** Lewis and Iglesias Consultants, Kingston, Ontario, Canada, October 1995.

Appendix A

Request from EDO for ACRS Review of the DPO

UNITED STATES
NUCLEAR REGULATORY COMMISSION
WASHINGTON, D.C. 20555-0001

July 20, 2000

MEMORANDUM TO: John Larkins
Executive Director
Advisory Committee on Reactor Safeguards

FROM: William D. Travers
Executive Director for Operations

SUBJECT: DIFFERING PROFESSIONAL OPINION ON STEAM GENERATOR
TUBE INTEGRITY ISSUES

The purpose of this memorandum is to request that the Advisory Committee on Reactor
Safeguards (ACRS) assist in the process to review a Differing Professional Opinion (DPO) on
Steam Generator Tube Integrity Issues. Specifically, I am requesting that the ACRS function as
the equivalent of an ad hoc panel, under Management Directive (MD) 10.159, to review the
DPO.

The issues raised in the DPO are reflected in the Staff Consideration Document dated
November 1, 1999, and the DPO Reply Document dated December 16, 1999 (and
attachments). Consideration of this differing professional opinion (DPO) has been proceeding
according to a memorandum dated December 29, 1998, included as an attachment, which
established a three-step approach. Step (1) publication of specific documents for public
comment, and Step (2) preparation of a final staff position, have been completed. The author
of the DPO, has completed his part of Step (3) by reviewing the staff's final position and
providing a response in which he identifies areas which he believes are still unresolved. The
appointment of an ad hoc panel to address the remaining issues completes Step (3). We have
attempted to establish an ad hoc panel comprised of members of the NRC staff who are
suitable for the task and acceptable to the DPO author. However, these attempts have been
unsuccessful. In light of the broad expertise and independence of the ACRS, I am requesting
that for this particular DPO, the ACRS function as the equivalent of an ad hoc panel described
in MD 10.159.

This DPO deals with complex technical issues. After completing the review, I request that the
ACRS provide me a summary report that documents its conclusions and any recommendations
relative to the pertinent technical issues.

Since 1991, an extensive record of documentation has been developed on the underlying
technical issues. These documents would be provided to the ACRS to assist in the review. To
facilitate transferring the collected documentation and information regarding the DPO, please
contact my staff to establish a mutually agreeable time to meet.

Thank you for your assistance in reviewing this important matter.

Appendix B

ACRS Acceptance of the EDO Request

UNITED STATES
NUCLEAR REGULATORY COMMISSION
ADVISORY COMMITTEE ON REACTOR SAFEGUARDS
WASHINGTON, D.C. 20555-0001

September 11, 2000

MEMORANDUM FOR: William D. Travers
                       Executive Director for Operations

FROM: D. A. Powers, Chairman, ACRS

SUBJECT: DIFFERING PROFESSIONAL OPINION ON STEAM
                       GENERATOR TUBE INTEGRITY ISSUES

In a memorandum dated July 20, 2000, to the ACRS Executive Director, you requested ACRS assistance in the technical resolution of a Differing Professional Opinion (DPO) associated with steam generator tube rupture events. Specifically, you requested that the ACRS function as the equivalent of an ad hoc panel, under Management Directive 10.159, to review the DPO on steam generator tube integrity issues, and provide you with a summary report documenting the conclusions and any recommendations relative to the pertinent technical issues. The ACRS has agreed to your request and plans to complete its review of the technical issues associated with the DPO by December 2000, barring unforseen circumstances.

SCOPE OF ACRS REVIEW

In addition to accepting your request, this memorandum attempts to clarify the scope of the ACRS review. We understand that the scope of the ACRS review is to assess the technical merits of the DPO issues and provide recommendations for your use in resolving the DPO. We assume that the main DPO issues, noted below, are accurately defined in the "Differing Professional Opinion Consideration Document," which is attached to your memorandum to Dr. Hopenfeld dated November 1, 1999.

- Nondestructive Examination (NDE) Issue
- Main Steam Line Break (MSLB) Issue
- Risk Increase Issue
- Iodine Spiking Issue
- Severe Accident Issues

Although the ACRS will focus on the issues in the DPO Consideration Document, dated September 22, 1999, and the DPO Authors Response to the EDO, dated January 5, 2000, there may be ancillary issues that the Committee may need to consider as part of its review. In performing this task, the Committee plans to review the referenced documents as well as other relevant documents.

During a meeting between the ACRS Executive Director and a member of his staff on July 24, 2000, Mr. Hopenfeld did not express concern regarding the ACRS serving as the ad hoc panel for reviewing the technical issues of his DPO except he expressed some concerns about previous ACRS decisions as noted in his memorandum to you dated July 28, 2000. We understand Dr. Hopenfeld's concerns about previous ACRS positions on these issues and we will attempt to minimize the influence of previous decisions in our review. The Committee will revisit its previous comments and recommendations on this matter included in the reports and letters listed below.

- ACRS report dated September 12, 1994, from T. S. Kress, ACRS Chairman, to Ivan Selin, NRC Chairman, Subject: Proposed Generic Letter 94-xx, "Voltage-Based Repair Criteria for Westinghouse Steam Generator Tubes."

- ACRS letter dated May 15, 1995, from T. S. Kress, ACRS Chairman, to James M. Taylor, EDO, Subject: Proposed Final Generic Letter 95-xx, "Voltage-Based Repair Criteria for Westinghouse Steam Generator Tubes."

- ACRS letter dated November 20, 1996, from T.S. Kress, ACRS Chairman, to James M. Taylor, EDO, Subject: Proposed Rule on Steam Generator Integrity.

- ACRS letter dated October 10, 1997, from R. L. Seale, ACRS Chairman, to L. Joseph Callan, EDO, Subject: Resolution of the Differing Professional Opinion Related to Steam Generator Tube Integrity.

PROPOSED REVIEW PROCESS

The Committee has established an Ad Hoc Subcommittee to review the technical merits of the DPO issues. The Subcommittee will function under the provisions of the Federal Advisory Committee Act (FACA). The Subcommittee and the full Committee will use the consultants, you have agreed to provide, and other consultants as needed to obtain technical support in reviewing certain DPO issues. After an initial meeting currently scheduled for October 10-13, 2000, the Subcommittee will decide on the scope and need for additional meetings. At the conclusion of the Subcommittee's review, the full Committee will discuss this matter and provide you with a letter, documenting its independent views on the DPO issues.

References:

1. Memorandum dated November 1, 1999, from William D. Travers, EDO, to Joram Hopenfeld, RES, Subject: Differing Professional Opinion on Steam Generator Tube Integrity Issues, with attachments:

   a. Differing Professional Opinion Consideration Document

   b. Public comments on Draft Regulatory Guide, DG-1074, "Steam Generator Tube Integrity."

2. Memorandum dated December 16, 1999, from Joram Hopenfeld, RES, to William D. Travers, EDO, Subject: Differing Professional Opinion on Steam Generator Tube

Integrity Issues (Response to the November 1, 1999 memorandum from the EDO) with attachments:

- a. Letter dated September 12, 1994, from T. S. Kress, Chairman, ACRS, to I. Selin, Chairman, NRC, Subject: Proposed Generic Letter 94-xx, "Voltage-Based Repair Criteria for Westinghouse Steam Generator Tubes."

- b. Letter dated October 21, 1997, from R. L. Seale, Chairman, ACRS, to S. A. Jackson, Chairman, NRC, "Summary Report - Four Hundred Fortieth Meeting of the Advisory Committee on Reactor Safeguards."

- c. J. Hopenfeld Comments on the Thermal Hydraulic Analysis in NUREG-1570, ACRS Materials and Metallurgy Subcommittee & Severe Accidents Subcommittee, March 5, 1997.

- d. Memoranda dated December 23, 1991 and March 27, 1992, regarding Differing Professional View.

- e. Memorandum dated September 11, 1992, from J. Hopenfeld to E. Beckjord, "Addendum to March 27, 1992, Memo Regarding Degraded Steam Generator Tubes."

- f. Memorandum dated September 28, 1999, from J. Hopenfeld to W. D. Travers, "DPO Panel Review of Steam Generator Integrity."

- g. J. Hopenfeld, "Differing Professional Opinion Regarding NRC Approach to Steam Generator Aging," September 25, 1998.

- h. Memorandum dated May 20, 1998 from J. Hopenfeld, RES, to J. T. Larkins, ACRS, "New Information Relative to Steam Generator Behavior During Severe Accidents."

- i. Memorandum dated July 13, 1994, from J. Hopenfeld, RES, to J. M. Taylor, EDO, "Differing Professional Opinion Regarding Voltage-Based Interim Repair Criteria for Steam Generator Tubes."

3. Memorandum dated April 5, 2000, from Joram Hopenfeld, RES, to William Travers, EDO, Subject: Supplement to My DPO Regarding Multiple Steam Generator Leakage (Originally filed as a DPV in December 1991 and filed as a DPO in July 1994).

4. Memorandum dated May 17, 2000, from Jack R. Strosnider, to James T. Wiggins, Subject: Issues Presented in Supplement to Differing Professional Opinion Regarding Steam Generator Tube Integrity.

NRC FORM 335
(2-89)
NRCM 1102,
3201, 3202

U.S. NUCLEAR REGULATORY COMMISSION

BIBLIOGRAPHIC DATA SHEET

(See instructions on the reverse)

1. REPORT NUMBER (Assigned by NRC, Add Vol., Supp., Rev., and Addendum Numbers, if any.)
NUREG-1740

2. TITLE AND SUBTITLE

Voltage-Based Alternative Repair Criteria

A Report to the Advisory Committee on Reactor Safeguards
by the Ad Hoc Subcommittee on a Differing Professional Opinion

3.	DATE REPORT PUBLISHED	
	MONTH	YEAR
	March	2001

4. FIN OR GRANT NUMBER

5. AUTHOR(S)

Ad Hoc Subcommittee on a Differing Professional Opinion

6. TYPE OF REPORT

Technical

7. PERIOD COVERED *(Inclusive Dates)*

8. PERFORMING ORGANIZATION - NAME AND ADDRESS *(If NRC, provide Division, Office or Region, U.S. Nuclear Regulatory Commission, and mailing address; if contractor, provide name and mailing address.)*

Ad Hoc Subcommittee on a Differing Professional Opinion

Advisory Committee on Reactor Safeguards

US Nuclear Regulatory Commission

Washington, DC 20555-0001

9. SPONSORING ORGANIZATION - NAME AND ADDRESS *(If NRC, type "Same as above"; if contractor, provide NRC Division, Office or Region, U.S. Nuclear Regulatory Commission, and mailing address.)*

Office of the Executive Director for Operations

US Nuclear Regulatory Commission

Washington, DC 20555-0001

10. SUPPLEMENTARY NOTES

11. ABSTRACT *(200 words or less)*

This report was prepared for the Advisory Committee on Reactor Safeguards (ACRS) as part of the Committee's effort to provide comments and recommendations to the Executive Director for Operations of the U. S. Nuclear Regulatory Commission for use in resolving a differing professional opinion (DPO) concerning voltage-based alternative criteria for the repair of flaws in steam generator tubes in pressurized water reactors. The report was prepared by an Ad Hoc Subcommittee of ACRS and its consultants. The report discusses the contentions that have been raised in the DPO and the staff responses to these contentions. Analyses and experimental results that support the various positions are described in summary fashion. Based on this information, the Subcommittee reaches a variety of conclusions and recommendations. The Subcommittee finds that alternative repair criteria are needed and that the general features of the criteria and the condition monitoring program that the staff has endorsed provide such alternative repair criteria that can adequately protect public health and safety. Analyses of the risk associated with adoption of the repair criteria need to better consider the progression of damage that can occur during design basis events, and especially the effects on tube integrity that may result from the dynamic processes associated with depressurization. The staff does not currently have a technically defensible analysis of how steam generator tubes, which may be flawed, will behave under severe accident conditions in which the reactor coolant system remains pressurized. Better databases are needed for the implementation of the condition monitoring of steam generators with 7/8" tubes. A program to detect systematic deviations from the bounding, linear model of flaw growth during operations is needed. The staff needs to develop a more technically defensible treatment of the iodine spiking phenomena associated with design basis events.

12. KEY WORDS/DESCRIPTORS *(List words or phrases that will assist researchers in locating the report.)*

Voltage-Based Alternative Criteria
Steam Generator Tubes
Pressurized Water Reactors
Design Basis Events
Iodine Spiking Phenomena
Differing Professional Opinion (DPO)

13. AVAILABILITY STATEMENT

Unlimited

14. SECURITY CLASSIFICATION

(This Page)

Unclassified

(This Report)

Unclassified

15. NUMBER OF PAGES

16. PRICE

NRC FORM 335 (2-89)

Federal Recycling Program

www.ingramcontent.com/pod-product-compliance
Lightning Source LLC
Chambersburg PA
CBHW081852170526
45167CB00007B/2989